Mesh Adaptation for Computational Fluid Dynamics 1

Mesh Adaptation for Computational Fluid Dynamics 1

Continuous Riemannian Metrics and Feature-based Adaptation

Alain Dervieux
Frédéric Alauzet
Adrien Loseille
Bruno Koobus

ISTE

WILEY

First published 2022 in Great Britain and the United States by ISTE Ltd and John Wiley & Sons, Inc.

ISTE Ltd
27-37 St George's Road
London SW19 4EU
UK

www.iste.co.uk

John Wiley & Sons, Inc.
111 River Street
Hoboken, NJ 07030
USA

www.wiley.com

Library of Congress Control Number: 2022934897

British Library Cataloguing-in-Publication Data
A CIP record for this book is available from the British Library
ISBN 978-1-78630-831-3

Contents

Acknowledgments

This book presents many theoretical and numerical accomplishments performed in collaboration with the following researchers:

Rémi Abgrall, Olivier Allain, Francoise Angrand, Paul Arminjon, Nicolas Barral, Anca Belme, Fayssal Benkhaldoun, Francois Beux, Gautier Brèthes, Véronique Billey, Alexandre Carabias, Romuald Carpentier, Giles Carré, Yves Coudière, Francois Courty, Didier Chargy, Paul-Henri Cournède, Christophe Debiez, Jean-Antoine Desideri, Gérard Fernandez, Loula Fezoui, Jérôme Francescatto, Loic Frazza, Pascal Frey, Paul-Louis George, Aurélien Goudjo, Nicolas Gourvitch, Damien Guégan, Hervé Guillard, Emmanuelle Itam, Marie-Hélène Lallemand, Stéphane Lanteri, Bernard Larrouturou, Anne-Cécile Lesage, David Leservoisier, Francoise Loriot, Mark Loriot, Laurent Loth, Nathalie Marco, Katherine Mer, Victorien Menier, Bijan Mohammadi, Eric Morano, Boniface Nkonga, Géraldine Olivier, Bernadette Palmerio, Gilbert Rogé, Bastien Sauvage, Éric Schall, Hervé Stève, Bruno Stoufflet, Francois Thomasset, Julien Vanharen, Ganesan Vijayasundaram, Cécile Viozat and Stephen Wornom; we also want to apologize to the people we forgot to mention.

Also we want to acknowledge our friends of INRIA and Lemma, and in particular Charles Leca, Olivier Allain, Nathalie and Philippe Boh, for their support. INRIA provided excellent conditions for research and writing of this book to the first three authors. Lemma permitted a rapid industrialization of our mesh adaptation methods.

The first author thanks his advisers, Jean Céa, Roland Glowinski and many thanks also to Charbel Farhat, Jacques Périaux and Roger Peyret.

This study is supported by fp6 and fp7 European progams (AEROSHAPE, HISAC, NODESIM, UMRIDA). The authors and their coworkers were granted access to the HPC resources of CINES/IDRIS under allocations made by GENCI (Grand Equipement National de Calcul Intensif).

Introduction

Numerical simulation is a central tool in the design of new human artifacts. This is particularly true in the present decades due to the difficult challenge of climate evolution. Yet recently climatic constraints were simply translated into the need for further progress in reducing pollution, a big job, in particular for specialists of numerical simulation. Today, it is likely that the use of numerical simulation, and particularly computational mechanics, will be central to the study of a new generation of human artifacts related to energy and transport. Fortunately, these new constraints are contemporary with the rise of a remarkable maturity of numerical simulation methods. One sign of this maturity is the flourishing of mesh adaptation. Indeed, mesh adaptation is now able to solve in a seamless way the deviation between theoretical physics and numerical physics, managed by the computer after discretization. A practical manifestation of this is that the engineer is freed from taking care of the mesh(es) needed for analysis and design. A second effect of mesh adaptation is an important reduction of energy consumption in computations, which will be amplified by the use of so-called higher order approximations. Mesh adaptation is thus the source of a new generation of more powerful numerical tools. This revolution will affect a generation of conceptualizers, numerical analysts and users, who are the engineers in design teams.

These books (Volumes 1 and 2) will be useful for researchers and engineers who work in computational mechanics, who deal with continuous media, and in particular who focus on computational fluid dynamics (CFD). They present novel mesh adaptation and mesh convergence methods developed over the last two decades, in part by the authors. They are expanded from a series of scientific articles, which are re-written, reorganized and completed in order to make the new content up-to-date, self-contained and more educational.

Let us describe in our own way the central role of meshes in the numerical simulation process, making it possible to compute a prediction of a physical phenomenon. In short, real-life mechanics consists of molecules and their interactions. The notion of continuous medium helps to transform a large but finitely complex system into a infinitely but smoothly complex one. For example, the understanding of gas flow relies today on the kinetic gas theory, which says that a gas is made up of a large number of molecules. We imagine the molecules as balls (monoatomic gas), but this is only a model for our imagination. We next consider that these balls are playing some sort of 3D "billiard" and that, if we are lucky, a continuum model describing the macroscopic behavior is a representative mean of the individual behaviors. In a similar manner, solids are made of a large number of molecules interacting with each other, and can be modeled as continuum media. Then the history of our billions of molecules is transformed into that of a continuous medium. Strictly speaking, the amount of information has gone from very large to infinitely large! But we have an extra assumption: that the infinitely complex functions that describe the continuous medium are smooth almost everywhere, because there exists a small scale such that even smaller scales behave in an expected way (predictable by interpolation, for example) except for some error that is smaller and smaller with the scale.

This assumption allows us many mathematical strategies:

– some laws for such functions can be constructed by considering that the functions have (regular enough) derivatives, with respect to time, and/or to space;

– such functions can be accurately represented by a set of special functions described by a small amount of information, such as polynomials, thanks, typically, to the Taylor formula.

The first point corresponds to the construction of partial differential equations (PDE). The second point corresponds the approximation or interpolation of known functions or the approximation of PDE solutions.

For clarity, we shall speak about "interpolation" for a discrete representation close to a known function and about "approximation" only for a discrete representation close to an (a priori unknown) PDE solution. Both strategies, interpolation and approximation, are related to *discretization*, the purpose of which is to reduce the infinitely complex continuous model into finitely complex discrete models, so that both the representation and the seeking of these for a particular physical situation is the matter of a finite amount of computational resources:

– a finite number of digits;

– a finite number of operations.

With respect to digits, any real number can still be extremely/infinitely complex and has to be replaced by a floating point representation. We shall not analyze the difference between a real number and its floating point representations, which is out of the scope of this book. For us, the only consequences of replacing real numbers by floating point representations is that using round-off truncation may amplify error modes in iterative algorithms. We shall have to choose only numerical algorithms that will produce good results even with round-off error, namely stable algorithms.

With respect to operations, our discrete model will rarely be computed on a sheet of paper but more frequently with a computer. It is common to call the complexity of an algorithm the number of elementary operations (additions, multiplications, etc.) necessary to perform it on a given set of data. We extend this notion by defining the complexity of the interpolation of a function as the number of numbers necessary to represent (with a given accuracy) that function. Two important ingredients of numerical analysis are connected by this notion:

– To *approximate a known function* with a certain level of accuracy, we need to handle (to store) some amount of real numbers, the degrees of freedom.

– To *compute* these degrees of freedom for an unknown function using an algorithm solving a system to provide the approximate solution, we need accordingly some amount of operations.

In both cases, we need to adapt a data structure to the geometrical domain on which the function has its definition set. This data structure, the mesh, generally locates nodes on the domain.

Approximating a smooth function can be done with a storage that increases at best only inverse-exponentially (spectral approximation) or, in most cases, inverse-polynomially (approximation of given order) with the prescribed error. The computation of this approximation can, at best, be done with linearly complex algorithms such as full multigrid. More precisely, let us consider the combination of

– a smooth unknown solution u of a PDE in a domain inside $I\!R^d$;

– an approximation of the PDE of order α for a certain norm $|.|$, that is,

$$|u - u_N| \leq K_1 N^{-\alpha/d}$$

with K_1 depending on the solution u and where N is the number of nodes of the computational domain; and

– a linearly complex solution algorithm, $CPU(N) = K_2 N$ with K_2 depending on the algorithm and where "CPU" is the computational time.

Then for a prescribed error $|u - u_N| \leq \varepsilon$, the CPU effort should be at least

$$CPU = K_2(\varepsilon/K_1)^{-d/\alpha}.$$

If the order α is one, the CPU increases like ε^{-3} in 3D (steady case) or even ε^{-4} in the unsteady 3D case. In many cases, this indicates that the computation at a useful accuracy level is simply not affordable. For a smooth function, the affordability increases importantly when order is increased.

In the case of a non-smooth function, the effort to represent it can be extremely large if this function involves a very large or infinite number of singularities. A more typical and interesting case is that of a function with a few singularities. To simplify our explanation, let us consider a Heavyside-type function, equal to 1 for an input larger than x_0, equal to -1 otherwise. This function is very easy to represent/approximate with a few (four) real numbers, but very complex to approximate with a series of smooth approximations lying on uniform meshes. Conversely, it can be accurately represented by smooth functions lying on an adapted mesh.

The object of these volumes is to present a few analyses and methods devoted to the relation between an approximation and its mesh as well as how to adapt the mesh to both the approximation and the precise computational case.

Let us review what is presented in the chapters of this volume.

Instead of presenting a mesh adaptation theory applying to an abstract family of approximation methods, we specify in Chapter 1 two particular approximation methods for compressible and two-fluid incompressible flows. We consider these particular approximation methods and try to cover a set of important questions to answer when we want to adapt a mesh to both the chosen approximation and the precise case to be computed.

The preliminary question is "why do we need to adapt the mesh?". Indeed, the natural approach is to use a uniform mesh. In order to understand how this will work, we need to study the convergence to the exact solution and the obstacles before reaching it (Chapter 2).

Second, adapting the mesh is searching for a good mesh, and why not the best mesh, based on some criteria of what constitutes the "best" mesh. Then a (mathematically) natural question is to ask in which set of meshes we should search the mesh. To answer this, we describe the continuous mesh representation, relying on Riemann metrics (Chapter 3).

Performing convergence consists of forcing an error to approach zero as much as we want. Equipped with the continuous mesh representation, we start with a rather simple error, the interpolation error of a *feature* of the flow, and look how to reduce it (Chapter 4). An essential ingredient is the rigourous setting of an optimization problem. We search a metric inside a subset of a Hilbert space, which minimizes a norm of the *interpolation error*. We are then equipped with a feature-based mesh adaptation method, which we call the *multiscale adaptation* when the norm chosen is a L^p norm with $1 < p < +\infty$, applying to steady mechanical problems (Chapters 5 and 6).

As a guide to the reader, an introduction to the basic methods can be obtained by reading Chapters 3–5, which are restricted to feature-based/multiscale adaptation for steady models. The sequel for steady models should concern goal-oriented methods, and the reader can directly pass to Chapter 6 of Volume 2.

In a first global reading of this volume, then, Chapters 2 and 6 can be passed over.

Numerical experiments described in the book are performed with three different computational codes with different features. They are shortly described in Chapter 1 and mentioned in each numerical presentation in order to fix the ideas concerning the important details of their implementation.

To make reading easier, many complex details are found in the annex of each chapter or, when not too long, in footnotes.

1

CFD Numerical Models

This chapter defines two fluid mechanics models and two particular numerical approximations for their discretization, which will be used in the examples of mesh adaptation algorithms that constitute the main part of this book. We first consider compressible fluid flows and introduce an approximation method, referred in the sequel as mixed element volume (MEV), which relies mainly on a standard continuous P_1-Galerkin approximation. Its stabilization is obtained by introducing high-order Godunov upwinding. Second, we consider a multifluid model based on the incompressible Navier–Stokes equations and introduce an approximation method based on the continuous P_1-Galerkin approximation. Pressure stabilization is obtained by projection. Advective stabilization is obtained by introducing high-order upwinding.

1.1. Compressible flow

1.1.1. *Introduction*

The simulation of compressible flows experienced, in the 1990s, a small revolution with the development of new algorithms that are able to compute flows through (or around) any kind of shape. This was due to new numerical algorithms and mesh generation algorithms. For both types, the main innovation was related to unstructured meshes, and the way to do it was first to rely on tetrahedrizations. Unstructured mesh generation and in particular tetrahedrization has been the object of many research and advances, and we refer, for example, to previous studies (Frey and George 2008; Borouchaki and George 2017; George et al. 2019, 2020) for monographies presenting these methods. The "any kind of shape" slogan has been progressively completed by the adaptation to any kind of flow, including flows in moving meshes, and by automatic mesh adaptation, appearing as an important issue to address in order to improve the expected benefits from a numerical simulation.

Several numerical methods have been developed for computing compressible flows on unstructured meshes. First, low-order methods (typically second-order) were developed. Let us mention central-differenced cell-centered and vertex-centered finite-volume methods (Jameson 1987; Mavriplis 1997), Taylor–Galerkin methods (Donea et al. 1987; Löhner et al. 1984), least–square Galerkin methods (Hughes and Mallet 1986) and the distributive schemes (Deconinck et al. 1993) for second-order accurate methods. Higher order accurate schemes have then been developed; let us mention unstructured ENO methods (Abgrall 1994) and discontinuous Galerkin methods (Cockburn 2003; Cockburn and Shu 1989).

In the first part of this section, we present and discuss a mixed finite-element/finite-volume low-order discretization for the Euler models of aerodynamics applicable to a very general class of tetrahedrizations, and we consider a few crucial numerical issues for the application of an Euler scheme:

– mastering numerical dissipation;

– mastering positiveness;

– evaluating the synergy between such kind of numerics and high-performance mesh adaptation methods.

In the second part, the extension of the MEV method to Navier–Stokes and Reynolds-averaged Navier–Stokes is considered.

The MEV discretization method is a combination of a finite-element method (FEM) with a vertex-centered finite-volume method (FVM). Like any vertex-centered approximation, it enjoys the property of handling the smallest number of unknowns for a given mesh and the possibility to assemble the fluxes on an edge-based mode. The underlying FEM is the standard Galerkin method with continuous piecewise linear approximation on triangles or tetrahedra. The FEM is applied directly for discretizing second-order derivatives (diffusion or viscosity terms). For hyperbolic terms, the FEM needs extra stabilization terms that are derived from an upwind FVM. The underlying FVM is a vertex-centered edge-based method. The finite-volume cell is built around each vertex, generally by using medians (2D) or median planes (3D); advection terms are stabilized with upwinding or artificial dissipation, and second-order "viscous" terms are discretized with finite elements. Among the different ways of constructing second-order accurate upwind schemes, the MUSCL formulation introduced by van Leer (1979) for finite-volume methods is particularly attractive and has been generally chosen.

The family of upwind MEV schemes was initiated by Baba and Tabata (1981) for first-order upwind diffusion-convection models and by Fezoui et al. for Euler flows (see Fezoui 1985; Dervieux 1985, 1987; Fezoui and Stoufflet 1989; Fezoui and Dervieux 1989; Stoufflet et al. 1996). It has been studied by many CFD teams (see, in particular, Whitaker et al. (1989); Anderson and Bonhaus (1994); Venkatakrishnan (1996); Barth (1994); Catalano (2002)). The framework proposed in Selmin and Formaggia (1998) can also be considered as an extension of MEV. Many developments and results relying on this family of schemes are regularly reported by Farhat and co-workers (Farhat and Lesoinne 2000). A particular advantage of MEV is its ability to perform well in combination with very irregular meshes. As a result, this scheme was identified as particularly convenient for developing methods for shape design (Farhat 1995; Nielsen and Anderson 2002; Vàzquez et al. 2004), for fluid–structure interaction with moving meshes (Farhat and Lesoinne 2000) and of course for anisotropic mesh adaptation (Loseille et al. 2007).

Several theoretical or methodological questions concerning MEV are addressed in this chapter:

– *Accuracy*. The basic scheme is introduced in section 1.1.2. In case of meshes with a bounded aspect ratio, the second-order accuracy of the underlying Galerkin method holds for steady-state problems, even for rather irregular meshes. For the unsteady case, since the mass matrix diagonalization is applied, the constraint on mesh regularity is somewhat stronger. But the behavior of the upwind versions of the MEV for highly stretched structured meshes is the main drawback of this class of schemes. Barth (1994) suggests a modification in the shape of finite-volume cells, which we describe in section 1.4.

– *Higher order*. Extension to second-order upwinding is based on a MUSCL formulation and is presented in section 1.1.3.

– *Superconvergent low dissipation versions*. In the case where the flow field under study is smooth, the numerical dissipation can be importantly reduced while not allowing Gibbs oscillations. In the linear theory, oscillations arise when high-frequency components of the solution are dispersed, that is, propagated with large phase velocity error, without enough dissipation to damp them. For reducing overall dissipation while avoiding oscillation, we follow the lines of higher order upwinding. This is obtained by introducing a new type of MUSCL reconstruction. Dissipation appears as relying on higher order even derivatives. Some versions of the new family show higher order convergence on regular or very smooth meshes. We call this property superconvergence. This method is presented in section 1.1.4.

– *Robustness and positivity*. In the 1980s, robustness of numerical schemes for hyperbolics was put in relation with positiveness, monotony and total variation diminishing properties. In section 1.1.6, we state several positivity results for MEV schemes.

– *Viscous flows*. The compressible study is completed in section 1.2 with the description of the numerical scheme for viscous and turbulent (statistical closure) flows.

1.1.2. *Spatial representation*

1.1.2.1. *Mathematical model*

We write the unsteady Euler equations as follows in the computational domain $\Omega \subset \mathbb{R}^3$:

$$\Psi(W) = \frac{\partial W}{\partial t} + \nabla \cdot \mathcal{F}(W) = 0 \quad \text{in } \Omega, \qquad [1.1]$$

where $W = {}^t(\rho, \rho u, \rho v, \rho w, \rho E)$ is the vector of conservative variables. $\mathcal{F}(W) = (\mathcal{F}_1(W), \mathcal{F}_2(W), \mathcal{F}_3(W))$ is the convective flux:

$$\mathcal{F}_1(W) = \begin{pmatrix} \rho u \\ \rho u^2 + p \\ \rho uv \\ \rho uw \\ (\rho E + p)u \end{pmatrix}, \mathcal{F}_2(W) = \begin{pmatrix} \rho v \\ \rho uv \\ \rho v^2 + p \\ \rho vw \\ (\rho E + p)v \end{pmatrix},$$

$$\mathcal{F}_3(W) = \begin{pmatrix} \rho w \\ \rho uw \\ \rho vw \\ \rho w^2 + p \\ (\rho E + p)w \end{pmatrix} \qquad [1.2]$$

so that the state equation becomes

$$\frac{\partial W}{\partial t} + \frac{\partial \mathcal{F}_1(W)}{\partial x} + \frac{\partial \mathcal{F}_2(W)}{\partial y} + \frac{\partial \mathcal{F}_3(W)}{\partial z} = 0.$$

Here, ρ, p and E represent, respectively, density, thermodynamical pressure and total energy per mass unit. Symbols u, v and w stand for the Cartesian components of velocity vector $\mathbf{u} = (u, v, w)$. For a calorically perfect gas, we have

$$p = (\gamma - 1)\big(\rho E - \frac{1}{2}\rho|\mathbf{u}|^2\big), \qquad [1.3]$$

where γ is constant. A weak formulation including boundary conditions of this system writes for $W \in V = \left[H^1(\Omega)\right]^5$ [1] as follows:

1 Space H^1 (respectively, H^k) consists of measurable functions with square integrable derivative up to order one (respectively, k).

$$\forall \phi \in V, \quad (\Psi(W), \phi) = \int_\Omega \left(\phi \frac{\partial W}{\partial t} + \phi \nabla \cdot \mathcal{F}(W) \right) \, d\Omega$$
$$- \int_\Gamma \phi \hat{\mathcal{F}}(W) \cdot \mathbf{n} \, d\Gamma = 0, \qquad [1.4]$$

where Γ is the boundary of the computational domain Ω (Figure 1.1), $\mathbf{n}$ is the outward normal to Γ and the boundary flux $\hat{\mathcal{F}}$ contains the boundary conditions detailed in viscous case in section 1.2.2. We are interested by this unsteady formulation together with the steady one, in which the time derivative is not introduced.

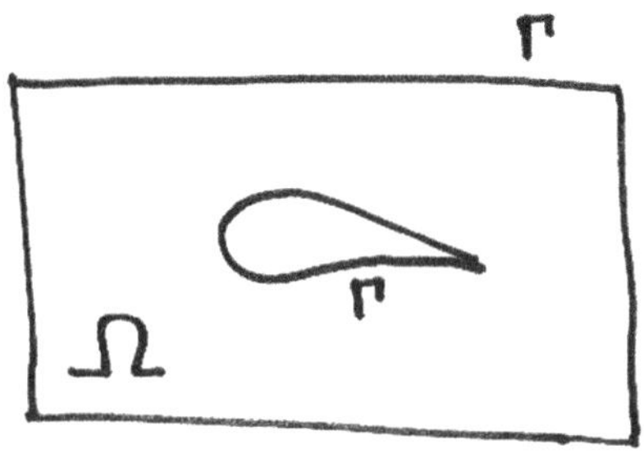

Figure 1.1. *A typical computational domain Ω limited by the two-component boundary $\Gamma = \partial\Omega$*

1.1.2.2. *Discrete variational representation*

We consider here the *steady* case, which is written as

$$\nabla \cdot \mathcal{F}(W) = 0$$

or in variational formulation:

$$\forall \phi \in V, \quad (\Psi(W), \phi) = \int_\Omega (\phi \nabla \cdot \mathcal{F}(W)) \, d\Omega - \int_\Gamma \phi \hat{\mathcal{F}}(W) \cdot \mathbf{n} \, d\Gamma = 0. \qquad [1.5]$$

The discretization chosen relies on two main choices. First, we consider a *tetrahedrization* as the discretization of the computational domain. This choice is made in connection with the progresses made for automatically generating and adapting meshes of this kind. Second, once the mesh is chosen, we have to put on it a set of nodes that are the geometrical supports of the degrees of freedom. The option chosen is the *set of vertices*. It is the option of the usual continuous P^1 FEM approximation. It corresponds to the smallest number of nodes for a given mesh. Let $\mathcal{T}_h$ be a tetrahedrization of Ω, which is admissible for finite elements, that is, Ω is partitioned in tetrahedra, and the intersection of two different tetrahedra is either empty, or a vertex, or an edge, or a face. The test functions are taken into the approximation space V_h made up of continuous piecewise linear functions included in $V = [H^1(\Omega)]^5$:

$$V_h \; = \; \left\{ \phi_h \;\middle|\; \phi_h \text{ is continuous and } \phi_{h|T} \text{ is linear } \forall T \in \mathcal{T}_h \right\}^5 .$$

In order to avoid the management of projectors applicable in the whole H^1 space, we work inside the following spaces:

$$\bar{V} = ([H^2(\Omega)]^5) \text{ and } \bar{V}_h = \bar{V} \cap V_h.$$

It is useful to introduce Π_h the corresponding P^1 interpolation operator:

$$\begin{aligned} \Pi_h : \bar{V}_h &\longrightarrow V_h \\ \phi &\longmapsto \Pi_h \phi \quad \text{with } \Pi_h \phi(i) = \phi(i) \;\; \forall i \text{ vertex of } \mathcal{T}_h. \end{aligned}$$

Then the discrete steady formulation of problem [1.5] is written as

$$\forall \phi_h \in V_h, \qquad \int_\Omega \phi_h \nabla \cdot \mathcal{F}_h(W_h) \, d\Omega - \int_\Gamma \phi_h \hat{\mathcal{F}}_h(W_h) \cdot \mathbf{n} \, d\Gamma \; = \; 0, \qquad [1.6]$$

where $\mathcal{F}_h$ is by definition the P^1 interpolate of $\mathcal{F}$ in the sense that

$$\mathcal{F}_h(W) = \Pi_h \mathcal{F}(W) \text{ and } \mathcal{F}_h(W_h) = \Pi_h \mathcal{F}(W_h), \qquad [1.7]$$

and, as the operator $\mathcal{F}_h$ applies to the values of W at the mesh vertices, we have

$$\mathcal{F}_h(W) = \mathcal{F}_h(\Pi_h W) = \Pi_h \mathcal{F}(\Pi_h W). \qquad [1.8]$$

We get the same relations for $\hat{\mathcal{F}}_h(W)$:

$$\hat{\mathcal{F}}_h(W) = \Pi_h \hat{\mathcal{F}}(\Pi_h W) \text{ and } \hat{\mathcal{F}}_h(W_h) = \Pi_h \hat{\mathcal{F}}(W_h). \qquad [1.9]$$

Practically, this definition means that nodal flux values are written as $\mathcal{F}_h(\mathbf{x}_i) = \mathcal{F}_h(W(\mathbf{x}_i))$, where fluxes $\mathcal{F}_h$ are evaluated at the mesh vertices i. Discrete fluxes functions $\mathbf{x} \mapsto \mathcal{F}_h(\mathbf{x})$ are derived from the nodal values by P^1 intrapolation inside every element. In contrast to the standard Galerkin approach, this definition emphasizes that the discrete fluxes are in V_h.

1.1.2.3. *MEV basic equivalence*

The discrete formulation [1.6] can be transformed into a *vertex-centered finite-volume scheme* applied to tetrahedral unstructured meshes. This assumes a particular partition in control cells C_i of the discretized domain Ω_h:

$$\Omega_h = \bigcup_{i=1}^{n_c} C_i \,, \qquad [1.10]$$

each control cell being associated with a vertex i of the mesh. The corresponding test functions are the piecewise constant characteristic functions of cells:

$$\chi^i(\boldsymbol{x}) = \begin{cases} 1 \text{ if } \boldsymbol{x} \in C_i, \\ 0 \text{ otherwise.} \end{cases}$$

Then, using the Stokes formula, the finite-volume weak formulation (steady case) becomes for each vertex i, that is, for each cell C_i,

$$\int_{\partial C_i} \mathbf{n}_i \mathcal{F}(W)\, d\sigma = 0, \qquad [1.11]$$

where $\mathbf{n}_i$ holds for the unit normal to ∂C_i outpointing from C_i.

DEFINITION 1.1.– *Median cell*: The *dual finite-volume cell* is built by the rule of medians. In 2D, the median cell is limited by segments of medians between centroids and mid-edge (Figure 1.2). In 3D, each tetrahedron T of the mesh is split into four hexahedra[2] constructed around each of its four vertices. For a vertex i, the hexahedron $C_i \cap T$ is defined by the following points (Figure 1.3):

i) the three middle points of the edges issued from i;

ii) the three gravity centers of the faces containing i;

iii) the center of gravity of the tetrahedron;

iv) the vertex i.

2 The quadrilateral between a mid edge, the two neighboring face centroids and tetrahedron's centroid being on a plane.

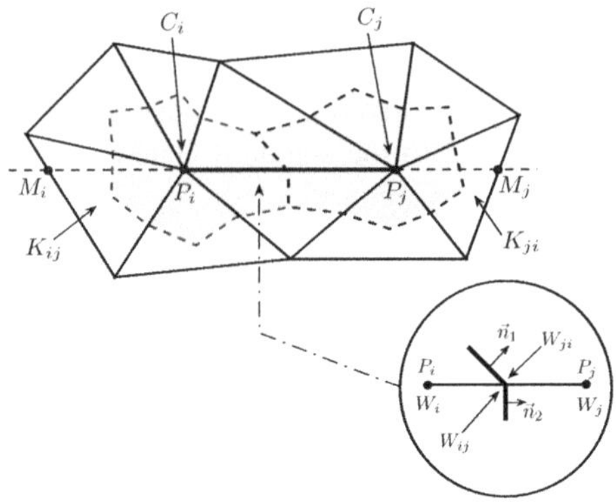

Figure 1.2. *Illustration of finite-volume cell construction in two dimensions with two neighboring cells, C_i and C_j, around i (P_i in the figure) and j (P_j in the figure), respectively, and of the upwind triangles K_{ij} and K_{ji} associated with the edge ij. Representation of the common boundary ∂C_{ij} with the solution extrapolated values for the MUSCL type approach. Dash lines are segments of medians of the triangles*

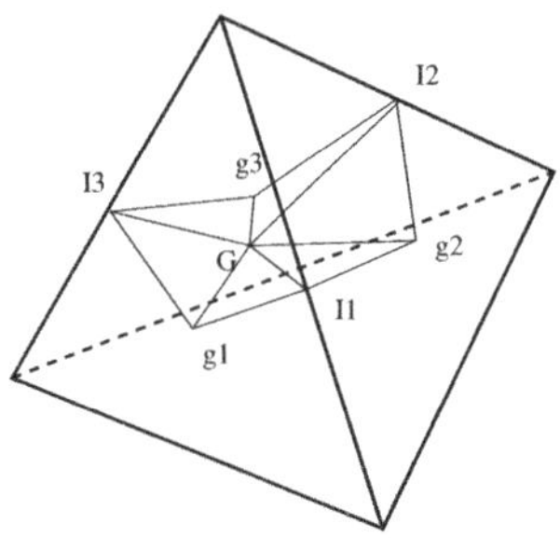

Figure 1.3. *The planes which delimit the finite-volume cell (related to upper vertex) inside a tetrahedron (3D case). G is the tetrahedron centroid, g_ks are face centroids and I_ks are edge centers*

The cell C_i of vertex i is the collection of all hexahedra linked to i. The common boundary $\partial C_{ij} = \partial C_i \cap \partial C_j$ between two neighboring cell C_i and C_j is divided into several triangular interface facets. □

An illustration of this construction is shown in Figure 1.4 for the 3D case.

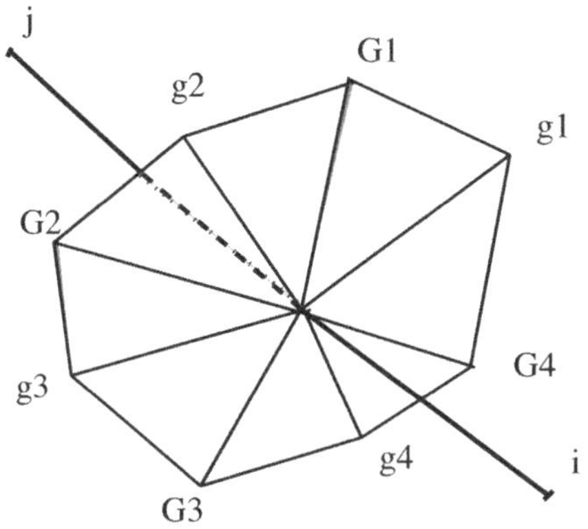

Figure 1.4. *Illustration of finite-volume cell interface ∂C_{ij} between two neighboring cells C_i and C_j (3D case)*

The finite-volume fluxes between cells around vertices i and j are integrated through the common boundary ∂C_{ij} with a value of $\mathcal{F}_h$ equal to the half-sum of $\mathcal{F}_h(W_i)$ and $\mathcal{F}_h(W_j)$:

$$\Phi_{ij}^{MEV} = \frac{\mathcal{F}_h(W_i) + \mathcal{F}_h(W_j)}{2}.\nu_{ij}, \qquad [1.12]$$

where ν_{ij} denotes the integral of the normal $\mathbf{n}_i$ to common boundary between cells C_i and C_j,

$$\nu_{ij} = \int_{\partial C_{ij}} \mathbf{n}_i \, d\sigma$$

and $W_i = W(i)$. The finite-volume formulation for an internal vertex i writes as the sum of all the fluxes evaluated from the vertices j belonging to $V(i)$ where $V(i)$ is the set of all neighboring vertices of i. Taking into account the boundary fluxes, the discrete scheme [1.6] then writes:

$$\sum_{j \in V(i)} \Phi_{ij}^{MEV} - \int_{\Gamma \cap \partial C_i} \bar{\mathcal{F}}_h(W_h) \cdot \mathbf{n} \, d\Gamma = 0. \qquad [1.13]$$

We obtain a vertex-centered finite-volume approximation, which is P^1-exact with respect to the flux function $\mathcal{F}_h$. This scheme enjoys most of the accuracy properties of the Galerkin method (Mer 1998), such as the second-order accuracy on any mesh for diffusion-convection models. However, it lacks stability and cannot be applied to purely hyperbolic models such as the Euler equations.

1.1.2.4. *Flux integration*

Once the cells are defined, the spatial divergence $div\mathcal{F}$ is transformed via the Stokes formula into integrals of normal fluxes $\mathcal{F}.\mathbf{n}$ at cell boundaries. In the proposed family of schemes, the accuracy of the integration quadrature on cell boundaries is not as crucial: we choose a very simple option, the *edge-based integration*. On the contrary, flux integration sets the important problem of scheme stabilization. The variables are assumed to be constant by cell, and therefore, they are discontinuous from a cell to its neighbor. Upwind integration will rely on the Godunov method based on the two different values at each side of the discontinuity.

1.1.2.4.1. Central differencing

Let us write a vertex-centered central differenced finite-volume scheme for the steady Euler equations applied to an unstructured mesh as follows:

$$\begin{aligned} \mathbf{\Psi}_h(\Gamma, W)_i &= 0, \quad \text{with} \\ \mathbf{\Psi}_h(\Gamma, W)_i &= \sum_{j \in V(i)} \Phi^{central}(W_i, W_j, \nu_{ij}) + \mathbf{B}_h(\Gamma, W)_i, \end{aligned} \qquad [1.14]$$

where $V(i)$ is the set of vertices that are neighbors of i, and ν_{jk} is the integral on interface between j and k of the normal vector. Symbol $\mathbf{B}_h(\Gamma, W)_i$ represents boundary fluxes in which Euler fluxes take into account the available boundary information. The centered integration for elementary flux Φ is written as follows:

$$\Phi^{central}(W_i, W_j, \nu_{ij}) = 0.5(\mathcal{F}_i + \mathcal{F}_j) \cdot \nu_{ij}, \qquad [1.15]$$

where $\mathcal{F}_i = \mathcal{F}(W_i)$ are the Euler fluxes computed at W_i. This is equivalent to introduce the following discrete space operator ∇_h^*:

$$\nabla_h^*(\mathcal{F})_i = \sum_{j \in V(i)} 0.5(\mathcal{F}_i + \mathcal{F}_j) \cdot \nu_{ij} \, / \, a(i), \qquad [1.16]$$

where $a(i)$ is the measure of cell C_i.

1.1.2.4.2. Godunov differencing

Godunov-type methods rely on the discontinuous representation of the unknowns at cell interfaces and on the computation of the fluxes at these discontinuities in function of both "left" and "right" values through the application of an approximate or an exact Riemann solver. This process introduces numerical viscosity terms that are very useful for stabilizing most flow calculations. First, we consider that W is constant by cell equal to W_i in C_i[3]. We write a vertex-centered first-order Godunov scheme for the Euler equations applied to an unstructured mesh as follows:

$$\mathbf{\Psi}_h(\Gamma, W)_i = \sum_{j \in V(i)} \Phi^{ARS}(W_i, W_j, \nu_{ij}) + \mathbf{B}_h(\Gamma, W)_i. \qquad [1.17]$$

Here, $\Phi^{ARS}(W_i, W_j, \nu_{ij})$ is evaluated by an approximate Riemann solver.

Roe approximate Riemann solver. A standard option is the Roe flux difference splitting (Roe 1981):

$$\Phi^{Roe}(W_i, W_j, \nu_{ij}) = 0.5(\mathcal{F}_i + \mathcal{F}_j) \cdot \nu_{ij} + 0.5|\mathcal{A}|(W_i - W_j) \qquad [1.18]$$

where $|\mathcal{A}|$ is the absolute value of the Jacobian flux along ν_{ij}:

$$\begin{aligned} \mathcal{A} &= \left(\frac{\partial \mathcal{F}}{\partial W}\right)_1 (\nu_{ij})_1 + \left(\frac{\partial \mathcal{F}}{\partial W}\right)_2 (\nu_{ij})_2 + \left(\frac{\partial \mathcal{F}}{\partial W}\right)_3 (\nu_{ij})_3 \quad \text{(3D case)} \\ \mathcal{A} &= T \Lambda T^{-1}, \quad \Lambda = \text{diagonal eigenvalues matrix}, \\ |\mathcal{A}| &= T |\Lambda| T^{-1}. \end{aligned} \qquad [1.19]$$

These matrices are computed at an intermediate value $\overline{W}_{ij}$ of W_i and W_j; in short, we have:

$$\overline{W}_{ij} = (\rho_i^{\frac{1}{2}} W_i + \rho_j^{\frac{1}{2}} W_j)/(\rho_i^{\frac{1}{2}} + \rho_j^{\frac{1}{2}})$$

which enjoys the following property:

$$\mathcal{F}(W_i) - \mathcal{F}(W_j) = \mathcal{A}(\overline{W}_{ij})(W_i - W_j).$$

In the fully supersonic cases where $\mathcal{A}(\overline{W}_{ij}) = |\mathcal{A}(\overline{W}_{ij})|$ or $\mathcal{A}(\overline{W}_{ij}) = -|\mathcal{A}(\overline{W}_{ij})|$, Roe's splitting is fully upwind. By the hyperbolicity

3 In the MUSCL method, we consider that the mean value in cell C_i is identical to the value at vertex i, an approximation bringing simplification but not permitting the extension to very high order.

assumption, matrix $\mathcal{A}(\overline{W}_{ij})$ can be diagonalized. The absolute value $|\mathcal{A}(\overline{W}_{ij})|$ is given as:

$$|\mathcal{A}(\overline{W}_{ij})| = TDiag(|\lambda_1|,|\lambda_2|,|\lambda_3|,|\lambda_4|,|\lambda_5|)T^{-1} = sign(\mathcal{A}(\overline{W}_{ij}))\mathcal{A}(\overline{W}_{ij}),$$

where $sign(\mathcal{A}) = TDiag(sign(\lambda_1), sign(\lambda_2), sign(\lambda_3), sign(\lambda_4), sign(\lambda_5))T^{-1}$. Thus, this averaging also permits the following equivalent formulation:

$$\begin{aligned}\Phi^{Roe}(W_i,W_j,\nu_{ij}) &= 0.5(\mathcal{F}_i+\mathcal{F}_j).\nu_{ij}\\ &\quad +0.5sign(\mathcal{A}(\overline{W}_{ij}))(\mathcal{F}(W_j)-\mathcal{F}(W_i)).\end{aligned} \quad [1.20]$$

HLLC approximate Riemann solver. The idea of the HLLC[4] solver (following Toro 1999) is to consider locally a simplified Riemann problem with two intermediate states depending on the local left and right states. The simplified solution to the Riemann problem consists of a contact wave with a velocity S_M and two acoustic waves, which may be either shocks or expansion fans. The acoustic waves have the smallest and the largest velocities (S_i and S_j, respectively) of all the waves present in the exact solution. If $S_i > 0$, then the flow is supersonic from left to right and the upwind flux is simply defined from $F(W_i)$ where W_i is the state to the left of the discontinuity. Similarly, if $S_j < 0$, then the flow is supersonic from right to left and the flux is defined from $F(W_j)$ where W_j is the state to the right of the discontinuity. In the more difficult subsonic case when $S_i < 0 < S_j$, we have to calculate $F(W_i^\star)$ or $F(W_j^\star)$. Consequently, the HLLC flux is given by:

$$\Phi_{ij}^{HLLC}(W_i,W_j,\mathbf{n}_{ij}) = \begin{cases} F(W_i)\cdot\mathbf{n}_{ij} & \text{if } S_i > 0\\ F(W_i^\star)\cdot\mathbf{n}_{ij} & \text{if } S_i \le 0 < S_M\\ F(W_j^\star)\cdot\mathbf{n}_{ij} & \text{if } S_M \le 0 \le S_j\\ F(W_j)\cdot\mathbf{n}_{ij} & \text{if } S_j < 0 \end{cases}. \quad [1.21]$$

where $W_i^\star$ and $W_j^\star$ are evaluated as follows. Let us denote $\eta = \mathbf{u}\cdot\mathbf{n}$. Assuming that $\eta^\star = \eta_i^\star = \eta_j^\star = S_M$, the following evaluations are proposed (Batten et al. 1997) (the subscripts $_i$ and $_j$ are omitted for clarity):

$$W^\star = \frac{1}{S-S_M}\begin{pmatrix}\rho(S-\eta)\\ \rho\mathbf{u}(S-\eta)+(p^\star-p)\mathbf{n}\\ \rho E(S-\eta)+p^\star S_M - p\eta\end{pmatrix} \text{ where } p^\star = \rho(S-\eta)(S_M-\eta)+p.$$

4 From Harten et al. (1983) for contact discontinuities.

A key feature of this solver is in the definition of the three waves velocity. For the contact wave, we consider:

$$S_M = \frac{\rho_j \eta_j (S_j - \eta_j) - \rho_i \eta_i (S_i - \eta_i) + p_i - p_j}{\rho_j (S_j - \eta_j) - \rho_i (S_i - \eta_i)},$$

and the acoustic wave speeds based on the Roe average (denoted by $^-$):

$$S_i = \min(\eta_i - c_i, \overline{\eta} - \overline{c}) \quad \text{and} \quad S_j = \max(\eta_j + c_j, \overline{\eta} + \overline{c}).$$

With such waves velocities, the approximate HLLC Riemann solver has the following properties (Toro 1999): it automatically (i) satisfies the entropy inequality, (ii) resolves isolated contacts exactly, (iii) resolves isolated shocks exactly and (iv) preserves positivity of ρ, T, p.

1.1.3. *Spatial second-order accuracy: MUSCL*

The above schemes with Roe or HLLC are spatially first-order accurate. First-order upwind schemes of Godunov type enjoy a lot of interesting qualities, and in particular HLLC enjoys formally monotonicity or, in the case of the Euler model ρ-, T- and p-positivity. They can be extended to second order by applying the MUSCL method. Indeed, the fact that the Godunov method builds fluxes between cells with unknown variables constant by cells implies first-order accuracy. van Leer (1979, 1977) proposed to reconstruct a linear interpolation of the variables inside each cell and then to introduce in the Riemann solver the boundary values of these interpolations. Further, the slopes used for linear reconstruction can be limited in order to represent the variable without introducing new extrema. The resulting MUSCL method produces positive second-order schemes. We describe now an extension of MUSCL to unstructured triangulations with dual cells. The MUSCL ideas also apply to reconstructions which are different on each interface between cells, or equivalently on each edge. Several slopes of a dependant variable F are defined on the two vertices i and j of an edge ij as follows:

1) *Gradients*. First, the *centered gradient* $(\nabla F)^c_{ij}$ is defined as $(\nabla F)^c_{ij} \cdot \vec{ij} = F_j - F_i$.

We consider a couple of two triangles inm and jrs (2D) or two tetrahedra $inmo$ and $jrst$ (3D), one having i as a vertex and the second having j as a vertex. With reference to Figure 1.5, we define ϵ_{in}, ϵ_{im} and ϵ_{io} (respectively, ϵ_{jr}, ϵ_{js} and ϵ_{jt}) as the components of vector $\vec{ji}$ (respectively, $\vec{ij}$) in the oblique system of axes $(\vec{in}, \vec{im}, \vec{io})$ (respectively, $(\vec{jr}, \vec{js}, \vec{jt})$):

$$\vec{ji} = \epsilon_{in}\,\vec{in} + \epsilon_{im}\,\vec{im}, + \epsilon_{io}\,\vec{io},$$

$$\vec{ij} = \epsilon_{jr}\,\vec{jr} + \epsilon_{js}\,\vec{js} + \epsilon_{js}\,\vec{js}.$$

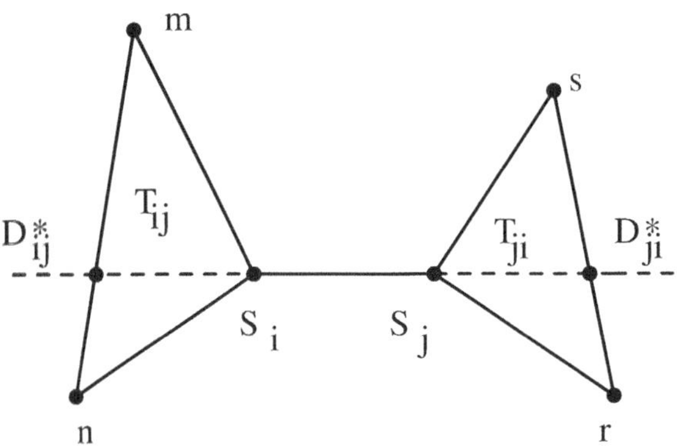

Figure 1.5. *Butterfly molecule in 2D: localization of the extra interpolation points D^*_{ij} and D^*_{ji} of nodal gradients. This allows to evaluate three derivatives along direction $ij = S_iS_j$, namely with D^*_{ij} and S_i, or S_i and S_j, or S_j and D^*_{ji}*

We say that T_{ij} and T_{ji} are upwind and downwind elements with respect to edge ij if the components $\epsilon_{in}, \epsilon_{im}, \epsilon_{io}, \epsilon_{jr}, \epsilon_{js}, \epsilon_{jt}$, are all non-negative:

$$T_{ij} \text{ upstream and } T_{ji} \text{ downstream } \Leftrightarrow Min(\epsilon_{in}, \epsilon_{im}, \epsilon_{io}, \epsilon_{jr}, \epsilon_{js}, \epsilon_{jt}) \geq 0. \quad [1.22]$$

The *upwind gradient* $(\nabla W)^u_{ij}$ is computed as the usual finite-element gradient on T_{ij} and the *downwind gradient* $(\nabla W)^d_{ij}$ on T_{ji}. This is written as:

$$(\nabla W)^u_{ij} = \nabla W\,|_{T_{ij}} \text{ and } (\nabla W)^d_{ij} = \nabla W\,|_{T_{ji}} \text{ where } \nabla W\,|_T = \sum_{k\in T} W_k \nabla\Phi_k|_T$$

are the P1-Galerkin gradients on triangle T.

2) *Interpolation at cell interface*. We now specify our method for computing the *interpolation slopes* $(\nabla W)_{ij}$ and $(\nabla W)_{ji}$:

$$(\nabla W)_{ij}\cdot\vec{ij} = (1-\beta)(\nabla W)^c_{ij}\cdot\vec{ij} + \beta(\nabla W)^u_{ij}\cdot\vec{ij}. \quad [1.23]$$

The computation of W_{ji} is analogous:

$$(\nabla W)_{ji} \cdot \vec{ij} = (1-\beta)(\nabla W)^c_{ij} \cdot \vec{ij} + \beta(\nabla W)^d_{ij} \cdot \vec{ij}. \qquad [1.24]$$

The coefficient β is an upwinding parameter that controls the combination of fully upwind and centered slopes and that is generally taken equal to $1/3$, according to the error analysis in Table 1.1.

Scheme	β	ξ^c	ξ^d	Spatial order	CFL_{max}
RK4(0.11, 0.2766, 0.5, 1)	1/3	0	0	3	1.9
RK6 (Abalakin et al. 2002a)	1/3	- 1/30	- 2/15	5	1.867
RK3-SSP	1/3	0	0	3	2

Table 1.1. *Superconvergent orders and maximal Courant numbers for MUSCL-third-order ($\xi^c = \xi^d = 0$) and V6 spatial schemes (1D analysis). The RK4 first and second coefficients are optimized for higher CFL with MUSCL*

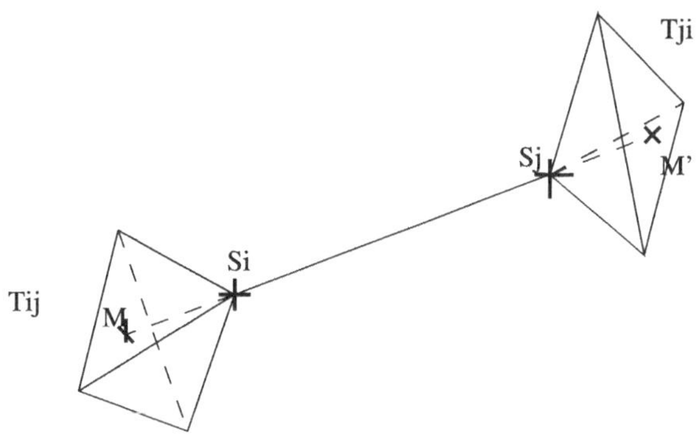

Figure 1.6. *Butterfly molecule in 3D: downwind and upwind tetrahedra are tetrahedra having, respectively, S_i and S_j as a vertex and such that line S_iS_j intersects the opposite face*

3) *Flux balance.* The scheme description is completed by replacing the first-order formulation [1.17] by the following flux balance:

$$\begin{aligned}&\mathbf{\Psi}_h(\Gamma, W)_i = \textstyle\sum_{j \in V(i)} \Phi^{ARS}(W_{ij}, W_{ji}, \nu_{ij}) \; + \; \mathbf{B}_h(\Gamma, W)_i \\ &\text{with :} \\ &W_{ij} = W_i + \tfrac{1}{2}(\nabla W)_{ij}.\vec{ij} \quad , \quad W_{ji} = W_j + \tfrac{1}{2}(\nabla W)_{ji}.\vec{ji}.\end{aligned} \qquad [1.25]$$

REMARK 1.1.– The numerical viscosity introduced by the MUSCL method is equivalent to a fourth-order derivative, weighted by a $O(h^3)$ coefficient, where h is local mesh size. When using certain very stretched meshes, the combination between MUSCL and median cells can produce a severe loss in accuracy. A cure for it is discussed in Annex 1 (section 1.4).

For this second-order version, the amount of dissipation, which is introduced, is the dominant term of the numerical error and may seem larger than needed in many applications.

1.1.4. *Low dissipation advection schemes*

The approximation described in the previous section is spatially second-order accurate. Note that the method combines finite differences in the local reconstruction and finite volume for fluxes. In contrast to reconstruction based on mean values (e.g. ENO schemes), a higher-order accurate interpolation does not bring a higher accuracy for the scheme. More precisely, for a nonlinear flux function, the accuracy of MUSCL schemes is limited to second order, as remarked by Wu and Wang (1995).

We now examine how to moderately change the reconstruction in order to improve the scheme. We get inspired by a direct simulation technique in which non-dissipative high-order approximations are stabilized in good accuracy conditions because of filters, which rely on very-high even order derivatives. In order to do this, we have to further extend the discretization stencil. Then it can be also interesting to choose a stencil extension, which also improves dispersion properties, since a less dispersive scheme needs less dissipation to avoid Gibbs-like oscillations. This leads to the idea of *superconvergent advection schemes* in the sense that they are of higher order (than two) on Cartesian mesh regions and for simpler models, like linear hyperbolic ones. More sophisticated versions giving high-order accuracy for nonlinear advection are described in Magoules (2011).

1.1.4.1. *Spatial scheme*

This section defines a low dissipation/low dispersion scheme introduced in Debiez (1996) and Debiez and Dervieux (1999). This scheme can be called *superconvergent* because it is designed in such a way that it is a higher-order scheme when applied to a Cartesian mesh. The scheme is built as follows:

1) A background flow $W = (\rho, \rho u, \rho v, E)$ on each vertex of the mesh is given.

2) Compute the primitive variable $\tilde{U} = (\rho, u, v, p)$ on each vertex (vertexwise loop).

3) Compute the nodal gradients $\nabla\tilde{U}$.

$$(\nabla\tilde{U})_i = \frac{1}{meas(C_i)} \sum_{T \in C_i} \frac{meas(T)}{3} \Big(\sum_{k \in T} (\tilde{U})_k \, \nabla \, \Phi_k|_T \Big) \quad \text{(2D case).} \qquad [1.26]$$

4) Start *edgewise assembly loop* by computing the extrapolated slopes:

$$\begin{aligned}(\nabla\tilde{U})_{ij} \cdot \vec{ij} = {} & (1-\beta)(\nabla\tilde{U})^c_{ij} \cdot \vec{ij} + \beta(\nabla\tilde{U})^d_{ij} \cdot \vec{ij} \\ & + \xi_c \left[(\nabla\tilde{U})^u_{ij} \cdot \vec{ij} - 2(\nabla\tilde{U})^c_{ij} \cdot \vec{ij} + (\nabla\tilde{U})^d_{ij} \cdot \vec{ij} \right] \\ & + \xi_d \left[(\nabla\tilde{U})_M \cdot \vec{ij} - 2(\nabla\tilde{U})_i \cdot \vec{ij} + (\nabla\tilde{U})_j \cdot \vec{ij} \right] \ ,\end{aligned} \qquad [1.27]$$

where $(\nabla\tilde{U})_M$ is the gradient at the point M, intersection of line ij with the face of T_{ij} which does not contain i as a vertex, as shown in Figure 1.6. The expression is analog for $\nabla(\tilde{U})_{ji}$. Let us define left and right variable interpolations:

$$\tilde{U}_{ij} \ = \ \tilde{U}_i + \nabla\tilde{U}_{ij} \quad ; \quad \tilde{U}_{ji} \ = \ \tilde{U}_j - \nabla\tilde{U}_{ji},$$

and recover the left and right values of conservative variables $W_{ij} = W(\tilde{U}_{ij})$, $W_{ji} = W(\tilde{U}_{ji})$. The upwind differenced flux is then written as follows (the index V6 holds for the numerical viscosity by sixth-order derivatives):

$$\boldsymbol{\Psi}_h^{V6}(\Gamma, W)_i \ = \ \sum_{j \in V(i)} \Phi^{ARS}(W_{ij}, W_{ji}, \nu_{ij}) \ + \ \mathbf{B}_h(\Gamma, W)_j. \qquad [1.28]$$

1.1.5. *Time advancing*

After semi-discretization of the unsteady Euler equation using the above scheme, by applying mass lumping we obtain

$$meas(C_i)\frac{dW_i}{dt} + \boldsymbol{\Psi}_h(\Gamma, W)_i = 0. \qquad [1.29]$$

The standard Runge–Kutta scheme can be used for time advancing the solution:

$$
\begin{aligned}
V_1 &= \Delta t\ \Psi(W^n)_i \\
V_2 &= \Delta t\ \Psi(W^n + V_1/2)_i \\
V_3 &= \Delta t\ \Psi(W^n + V_2/2)_i \\
V_4 &= \Delta t\ \Psi(W^n + V_3)_i \\
W_i^{n+1} &= W_i^n + V_1/6 + V_2/3 + V_3/3 + V_4/6,
\end{aligned}
\tag{1.30}
$$

where $\Psi(W)_i = -\mathbf{\Psi}_h(\Gamma, W)_i/meas(C_i)$.

In many case, a linearized version of the time stepping can be used. Jameson (1993) writes as follows (N-stage version):

$$
\begin{aligned}
W^{(0)} &= W^n \\
W^{(k)} &= W^{(0)} + \frac{\Delta t}{N-k+1}\Psi\left(W^{(k-1)}\right), \quad k = 1 \ldots N \\
W^{n+1} &= W^{(N)}.
\end{aligned}
\tag{1.31}
$$

An A-stability analysis as in Hirsch (1988) can be applied. We give in Table 1.1 some typical maximal CFL numbers for the six-stage RK scheme, which ensure a global accuracy order of five for the two best schemes of the proposed family. The RK3-SSP is written as

$$U^{(1)} = U^n + \Delta t\, L(U^n), \tag{1.32}$$

$$U^{(2)} = \frac{3}{4}U^n + \frac{1}{4}U^{(1)} + \frac{1}{4}\Delta t\, L(U^{(1)}), \tag{1.33}$$

$$U^{n+1} = \frac{1}{3}U^n + \frac{2}{3}U^{(2)} + \frac{2}{3}\Delta t\, L(U^{(2)}) \tag{1.34}$$

(see Shu and Osher 1988; Spiteri and Ruuth 2002). Table 1.1 illustrates that the above schemes can be used with CFL number of the order of unity. In the sequel, we consider spatial improvements for monotony and positivity. When using these improvements, the explicit time advancing to be used is the RK3-SSP.

All of the above spatial schemes can be advanced in time with implicit schemes such as BDF1 and BDF2. Details of implementation for the complete viscous model are given in section 1.2.11. A spatially first-order accurate simplified Jacobian is systematically used. In the unsteady case, that option is used inside a two-step

Newton-like process, referred as unsteady defect correction (Martin and Guillard 1996). Linear stability is unconditional in all cases. In the case where we seek a steady solution or a slowly evoluting solution during a long time, the efficiency of an explicit scheme applying on unstructured mesh is severely limited by the Courant condition on the time step. The implicit time advancing is mandatory and needs to be combined with an efficient linear solver.

REMARK 1.2.– It can be efficient to apply a multigrid iteration in combination with pseudo-time advancing (steady case) or (for both cases) an efficient implicit time advancing. Designing a multigrid scheme for unstructured meshes rises the problem of defining a series of coarser grids. In other words, we have to define several new meshes or to find an alternative strategy. In Francescatto and Dervieux (1998) and Lallemand et al. (1992), this is done in a transparent manner from the fine mesh by using the so-called cell agglomeration. Parallel multigrid extensions are proposed in Mavriplis (1997) and Fournier et al. (1998). □

Another option which is well adapted to message passing parallelism is a Krylov–Newton–Schwarz (KNS) algorithm, as in Knoll and Keyes (2004). A first version of KNS, under the form of the restrictive additive Schwarz (RAS), was developed in Cai and Sarkis (1999) and Sarkis and Koobus (2000). A deflation-based coarse grid extension of RAS is studied in Alcin et al. (2013).

The *Code Aironum* of INRIA and Montpellier University, derived from the AERO software developed in a collaboration between University of Colorado at Boulder and INRIA, involves all the above features (in particular RAS) in a massively parallel implementation.

1.1.5.1. *Conclusion on the superconvergent scheme*

Based on MUSCL schemes, the V6 schemes involve sophisticated primitive variable reconstruction designed in order to enjoy low dissipation properties because of a model of sixth derivative. In the case of linear advection with uniform velocity, the scheme presents superconvergence properties on Cartesian subregions of mesh where convergence is close to the fifth order. On unstructured non-Cartesian subregions (where limiters do not apply, see the next section), convergence reduces to an order between the second order and fifth order. Interested readers are referred to some previous studies (Debiez 1996; Debiez and Dervieux 1999; Debiez et al. 1998; Abalakin et al. 2001, 2002a,b,c, 2004; Camarri et al. 2001, 2002, 2004; Gourvitch et al. 2004; Abalakin and Kozubskaya 2014).

1.1.6. *Positivity of mixed element-volume formulations*

1.1.6.1. *Introduction*

Some of the most useful properties of upwind approximation schemes for hyperbolic equations are their monotonicity and positivity properties. For example, for the Euler equations, and by combining flux splitting and limiters, Perthame and co-workers (Perthame and Khobalate 1992; Perthame and Shu 1996) proposed second-order accurate schemes that maintain a density and a temperature positive. We refer also to Linde and Roe (1998) and the workshop by Venkatakrisnan (1998). Non-oscillating schemes (Harten and Osher 1987; Cockburn and Shu 1989) propose high accuracy approximations applicable to many stiff problems. However, they do not enjoy a strict satisfaction of positivity or monotony, which remains an important issue for stiff simulations, particularly in relation with highly heterogeneous fluid flows (see, for example, Abgrall 1996 or Murrone and Guillard 2005).

The present section defines positive/monotone versions of the MEV scheme. We first examine the maximum principle for a scalar conservation law, then we introduce a flux splitting that preserves density positivity for the Euler equations in 1D. This provides a basis to construct *multidimensional* schemes, applying to unstructured meshes and ensuring density positivity and maximum principle for convected species. This is of paramount importance for most flows of industrial interest for two reasons. The first one is that many industrial flows are at medium Mach number and are rather stable when negative densities are avoided. The second reason is that in many Reynolds-averaged Navier–Stokes flows, limiters are not necessary for the mean flow itself, but robustness problems arise in the computation of turbulence closure variables such as k and ε.

1.1.6.2. *Positive schemes and LED schemes for nonlinear scalar conservation laws*

This section recalls some basic properties. We keep the usual TVD/LED (TVD: total variation diminishing, LED: local extremum diminishing) criterion preferably to weaker positivity criteria in order to deal also with the maximum principle. We consider a nonlinear *scalar* conservation law form for the unknown $U(\mathbf{x}, t)$:

$$\frac{\partial U}{\partial t}(\mathbf{x}, t) + \nabla_{\mathbf{x}} \cdot \mathcal{F}\Big(U(\mathbf{x}, t)\Big) = 0, \qquad [1.35]$$

where t denotes the time and $\mathbf{x}$ denotes the space coordinate. For the 1D (nonlinear) case and under some classical assumptions (Godlewski and Raviart 1996), the solution

U satisfies a *maximum principle* that, for simplicity, we write in the case of the whole space:

$$\min_{\mathbf{x}} U(\mathbf{x}, 0) \leq U(\mathbf{x}, t) \leq \max_{\mathbf{x}} U(\mathbf{x}, 0). \qquad [1.36]$$

In 2D and 3D, but for the advection model, a positivity principle also holds. In the general multidimensional case, no TVD property applies. But the TVD numerical method can still be introduced in order to build non-oscillatory schemes. Note, however, that for compressible Euler and Navier–Stokes, the positivity of density, pressure, temperature, entropy, etc., holds in the continuous case and can be partly recovered in the discrete case by using TVD/LED numerics. From a positivity standpoint, it is enough to adopt the weaker LED criterion, which we do in next section.

Assuming that the mesh nodes are numbered, we call U_i the value at mesh node i. We recall now the classical positivity statement for an explicit time-integration (the proof is immediate).

LEMMA 1.1.– *A positivity criterion.* Suppose that an explicit first-order time-integration of equation [1.35] can be expressed under the form

$$\frac{U_i^{n+1} - U_i^n}{\Delta t} = b_{ii} U_i^n + \sum_{j \neq i} b_{ij} U_j^n \,, \qquad [1.37]$$

where all the b_{ij}, $j \neq i$, are non-negative and $b_{ii} \in I\!R$. Then it can be shown that the above explicit scheme preserves positivity under the following condition on the time-step Δt: $b_{ii} + \frac{1}{\Delta t} \geq 0$.

The maximum principle can be evaluated with the incremental form introduced by Harten (1983) and used by Jameson (1993) as follows for defining multidimensional LED schemes (see also Godlewski and Raviart 1996):

LEMMA 1.2.– *A LED criterion (Jameson 1993).* Suppose that an explicit first-order time-integration of equation [1.35] can be written in the form

$$\frac{U_i^{n+1} - U_i^n}{\Delta t} = \sum_{k \in V(i)} c_{ik}(U^n)\,(U_k^n - U_i^n), \qquad [1.38]$$

with all $c_{ik}(U^n) \geq 0$, and where $V(i)$ denotes the set of the neighbors of node i. Then the previous scheme verifies that a local maximum cannot increase and a local minimum cannot decrease, and under an appropriate condition on the time-step the positivity and the maximum principle are preserved[5].

1.1.6.2.1. First-order space-accurate MEV schemes

We integrate [1.35] over a cell C_i, integrating by parts the resulting convective fluxes and using a conservative approximation leads to the following semi-discretization of [1.35]:

$$a_i \frac{dU_i}{dt} + \sum_{j \in V(i)} \Phi(U_i, U_j, \nu_{ij}) = 0\,, \qquad [1.39]$$

where a_i is the measure of cell C_i. In the above semi-discretization, the values U_i and U_j correspond to a constant per cell *interpolation* of the variable U, and Φ is a *numerical flux function* so that $\Phi(U_i, U_j, \nu_{ij})$ approximates $\int_{\partial C_{ij}} \mathcal{F}(U) \cdot \mathbf{n}_i(\sigma) d\sigma$. In general, the numerical flux function $\Phi : (u, v, \nu) \mapsto \Phi(u, v, \nu)$ is assumed to be Lipschitz continuous, monotone increasing with respect to u, monotone decreasing with respect to v, and consistent:

$$\Phi(u, u, \nu) = \mathcal{F}(u) \cdot \nu. \qquad [1.40]$$

As already noted, taking $\Phi = \Phi^{HLLC}$ [1.21] makes this discretization a ρ, T, p-positive one under a usual time step condition.

1.1.6.2.2. Three-entry limited high-order space-accurate MEV scheme

We replace the values U_i and U_j by "better" interpolations U_{ij} and U_{ji} at the interface ∂C_{ij}. More precisely, the first-order MEV scheme becomes:

$$a_i \frac{dU_i}{dt} + \sum_{j \in V(i)} \Phi(U_{ij}, U_{ji}, \nu_{ij}) = 0, \qquad [1.41]$$

where U_{ij} and U_{ji} are left and right values of U at the interface ∂C_{ij}. Our purpose is to build a scheme that is non-oscillatory and positive. It has been early proved

5 PROOF.– Given U_i^n a local maximum, we deduce that $(U_k^n - U_i^n) \leq 0$ for all $k \in V(i)$. Therefore, [1.38] implies that $\dfrac{U_i^{n+1} - U_i^n}{\Delta t} \leq 0$, and the local maximum cannot increase. Likewise, we can prove that a local minimum cannot decrease. On the other hand, the reader can easily check that the positivity and the maximum principle are preserved when the time-step satisfies $\dfrac{1}{\Delta t} - \sum_{k \in V(i)} c_{ik}(U^n) \geq 0$. □

that high-order positive schemes must be necessarily built with a nonlinear process, the *limiter*. A limiter locally chooses between the first-order monotone version of the schemes and a higher-order one. In the case of unstructured meshes and scalar models, second-order positive schemes were derived using a two-entry symmetric limiter by Jameson (1993). Here, instead of the Jameson symmetric limiter, we work with the MUSCL formulation, involving two limiters per edge. The adaptation to triangulations (or tetrahedrizations) is close to the one proposed in Fezoui and Dervieux (1989) and Stoufflet et al. (1996). Following Debiez (1996), we extend it to a three-entry limiter that allows to design a positive scheme of third- (or even fifth-)order far from extrema when U varies smoothly. Then [1.41] becomes

$$a_i \frac{dU_i}{dt} + \sum_{j \in V(i)} \Phi(U_i + \frac{1}{2} L_{ij}(U), U_j - \frac{1}{2} L_{ji}(U), \nu_{ij}) = 0. \qquad [1.42]$$

In order to define $L_{ij}(U)$ and $L_{ji}(U)$, we use the upstream and downstream triangles (or tetrahedra) T_{ij} and T_{ji} (see Figures 1.6 and 1.7), as introduced in Fezoui and Dervieux (1989). Element T_{ij} is *upstream* to vertex i with respect to edge ij if for any sufficiently small real number η, the vector $-\eta\vec{ij}$ is inside element T_{ij}. Symmetrically, element T_{ji} is *downstream* to vertex i with respect to edge ij if for any small enough real number η the vector $-\eta\vec{ji}$ is inside element T_{ji}. If the 3D case is now considered, let i, n, m, o (respectively, j, r, s, t) be the four vertices of tetrahedron T_{ij} (respectively, T_{ji}), and let ϵ_{in}, ϵ_{im}, ϵ_{io} (respectively, ϵ_{jr}, ϵ_{js}, ϵ_{jt}) be the components of vector $\vec{ji}$ (respectively, $\vec{ij}$) in the oblique system of axes $(\vec{in}, \vec{im}, \vec{io})$ (respectively, $(\vec{jr}, \vec{js}, \vec{jt})$), we have

$$\vec{ji} = \epsilon_{in}\,\vec{in} + \epsilon_{im}\,\vec{im} + \epsilon_{io}\,\vec{io}, \quad ; \quad \vec{ij} = \epsilon_{jr}\,\vec{jr} + \epsilon_{js}\,\vec{js} + \epsilon_{jt}\,\vec{jt}.$$

Then T_{ij} and T_{ji} are upstream and downstrean elements, which mean that they have been chosen in such a way that the components ϵ_{in}, etc., are all non-negative, according to [1.22]. Let us introduce the following notations:

$$\Delta^- U_{ij} = \nabla U\,|_{T_{ij}} . \vec{ij}, \;\; \Delta^0 U_{ij} = U_j - U_i \;\; \text{and} \;\; \Delta^- U_{ji} = \nabla U\,|_{T_{ji}} \cdot \vec{ij},$$

where the gradients are those of the P1 (continuous and linear) interpolation of U. Jameson (1993) has noted that (for the 3D case):

$$\Delta^- U_{ij} = \epsilon_{in}\,(U_i - U_n) + \epsilon_{im}\,(U_i - U_m) + \epsilon_{io}\,(U_i - U_o),$$

and

$$\Delta^- U_{ji} = \epsilon_{jr}\,(U_r - U_j) + \epsilon_{js}\,(U_s - U_j) + \epsilon_{jt}\,(U_t - U_j),$$

with the same non-negative ϵ_{in}, ϵ_{im}, ϵ_{io}, ϵ_{jr}, ϵ_{js} and ϵ_{jt}.

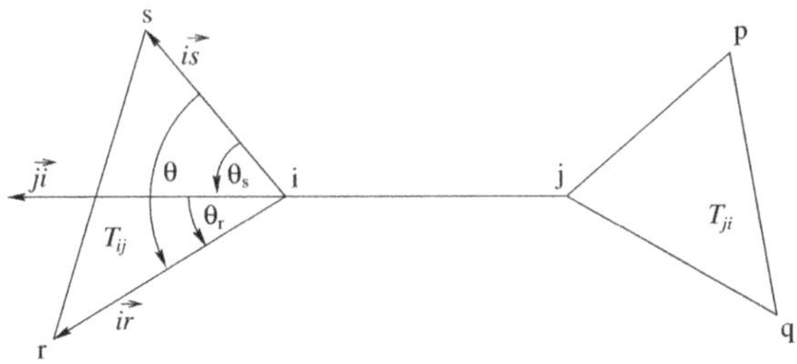

Figure 1.7. *Two-dimensional case: splitting of vector $\vec{ji}$ in the directions of downstream and upstream triangles edges*

Now, we introduce a family of continuous limiters with three entries, satisfying:

– $(P1)$ $L(u,v,w) \; = \; L(v,u,w)$

– $(P2)$ $L(\alpha\, u, \alpha\, v, \alpha\, w) \; = \; \alpha\, L(u,v,w)$

– $(P3)$ $L(u,u,u) \; = \; u$

– $(P4)$ $L(u,v,w) \; = \; 0$ if $uv \leq 0$

– $(P5)$ $0 \leq \frac{L(u,v,w)}{v} \leq 2$ if $v \neq 0$.

Note that there exists K^- and K^0 depending on (u,v,w) such that

$$L(u,v,w) = K^- u = K^0 v\,, \quad \text{with} \quad 0 \leq K^- \leq 2\,,\ 0 \leq K^0 \leq 2. \qquad [1.43]$$

A function verifying (P1) to (P5) exists, and in the numerical examples, we shall use the following version of the Superbee method of Roe:

$$\begin{cases} L_{SB}(u,v,w) = 0 \qquad \text{if} \quad uv \leq 0 \\ \qquad\qquad = Sign(u)\min(\,2\,|u|,\,2\,|v|,\,|w|)\ \text{otherwise.} \end{cases} \qquad [1.44]$$

We define

$$L_{ij}(U) = L(\Delta^- U_{ij},\, \Delta^0 U_{ij},\, \Delta^{HO} U_{ij}) \qquad [1.45]$$

$$L_{ji}(U) = L(\Delta^- U_{ji},\, \Delta^0 U_{ij},\, \Delta^{HO} U_{ji}), \qquad [1.46]$$

where $\Delta^{HO} U_{ji}$ is a third way of evaluating the variation of U which we can introduce for increasing the accuracy of the resulting scheme (see the following remark). Under the above assumptions, the LED property can be established and in particular, the positivity of density for Euler discretizations:

LEMMA 1.3.– The scheme defined by conditions P1-P4 and equations [1.45] and [1.46], based on a ρ-positive flux splitting is ρ-positive under a CFL-like condition.

PROOF.– See Cournède et al. (2006). □

REMARK 1.3.– Let assume that the option $L_{ij}(U) = \Delta^{HO} U_{ij}$ gives a high-order approximation with order ω. For U to be sufficiently smooth and assuming that the mesh size is smaller than α (small), there exists $\epsilon(\alpha)$ such that if $|\nabla U.\vec{ij}|/||\vec{ij}|| > \epsilon(\alpha)$ for an edge ij, the limiter L_{ij} is not active, that is,$L_{ij}(U) = \Delta^{HO} U_{ij}$. Then, the scheme is locally of order ω. □

REMARK 1.4.– In the case of a viscous flow, a particular attention has to be paid to the discrete diffusion operators that should also preserve positivity of variables; for finite-element discretization, this is related to standard acute angle condition (see Baba and Tabata 1981).

The proposed analysis does not need any assumption concerning the terms $\Delta^{HO} U_{ij}$. One option is the *superconvergent high order*, which allows computing a flow with, at the same time, lower dissipation and shock capturing. For $\Delta^{HO} U_{ij}$ we recommend using the extrapolated slopes defined in [1.27]. This option is useful for unsteady flows. It should be noted that dissipation can be even further decreased by adding sensors dedicated to the inhibition of limiters in regions where the flow is regular. Limiters should be avoided or locally well controlled in calculations, which need a very low level of dissipation, such as large eddy simulation and aeroacoustics. A second option is the usual *third-order*, which relies on the following flux:

$$\Delta^{HO3} U_{ij} = \frac{1}{3}\,\Delta^{-} U_{ij} + \frac{2}{3}\,\Delta^{0} U_{ij}, \qquad [1.47]$$

which gives us a third-order space-accurate scheme for linear advection on Cartesian triangular meshes. Since $\Delta^{HO3} U_{ij}$ is expressed from the two first entries, it produces in fact a "two-entry" limiter.

We propose now the analysis of these two-entry quasi-third order limited schemes with the important purpose of obtaining converged and accurate steady solutions. Let us define

$$R = \Delta^{0} U_{ij} / \Delta^{-} U_{ij}.$$

The extrapolation to mid-edge $U_{ij} = U_i + \frac{1}{2}L_{ij}(U)$ (see [1.42]) is written as

$$U_{ij} = U_i + \frac{1}{2}\psi(R)\Delta^- U_{ij} \quad \text{or} \quad U_{ij} = U_i + \frac{1}{2}\phi(\Delta^0 U_{ij}, \Delta^- U_{ij}).$$

The above limiter reduces to the one introduced by Koren (1993):

$$\psi_{Ko}(R) = min(1, 2R, \frac{1}{3} + \frac{2}{3}R) \quad \text{or} \quad \phi_{Ko}(a,b) = min(2a, 2b, \frac{1}{3}b + \frac{2}{3}a),$$

where ψ and ϕ are zero for $R \leq 0$. It is well known that this limiter shows a compressive behavior, with stiffer shocks. Also, it presents discontinuous derivatives. These properties have been identified as the cause of difficulties in convergence to steady state.

1.1.6.2.3. Piperno-limited MEV scheme

Piperno (1996) and Piperno and Depeyre (1998) have defined five design criteria for a *smooth limiter* based on the above third-order formulation:

i) the higher order gradient is taken for $R = 1$. Indeed, this is in the smooth part and we want to maximize the accuracy;

ii) function ψ has to be C^1;

iii) since the upwind scheme is fully second-order and does not oscillate for $R \to \infty$, let us have $\psi(R) \to 1$ when $R \to \infty$;

iv) moreover, the curve should be flat when $R \to \infty$, that is, $\psi'(R) \to 0$ when $R \to \infty$;

v) at $R = 1$, we should have at least C^2, and possibly the curve should coincide with the order three, that is, $\psi(R) = \frac{1}{3} + \frac{2}{3}R$ in a large interval around $R = 1$, in order to preserve third-order in this region.

Taking into account these criteria, Piperno defined the following limiter:

$$\psi_{Piperno}(R) = (\frac{1}{3} + \frac{2}{3}R)\begin{cases} \dfrac{3\frac{1}{R^2} - 6\frac{1}{R} + 19}{\frac{1}{R^3} - 3\frac{1}{R} + 18} & \text{if } R < \frac{1}{2} \\ 1 + (\dfrac{3}{2}\dfrac{1}{R} + 1)(\dfrac{1}{R} - 1)^3 & \text{if } R \geq 1. \end{cases}$$

This limiter is smoother than the previous one and is preferred for computing steady flows.

1.1.6.2.4. Gamma limiter

We follow an idea of Koren (1993) and enlarge the order three interval (criterion (v)) to have it between $R = 1/k$ and $R = k$, for $k \geq 2$. The novel limiter is written as

$$\psi_\gamma(R) = (\frac{1}{3} + \frac{2}{3}R) \begin{cases} \dfrac{3\frac{1}{R^2} - 6\frac{1}{R} + 19}{(\frac{1}{R} - k)^3 + 3\frac{1}{R^2} - 6\frac{1}{R} + 19} & \text{if } R < \frac{1}{k} \\ 1 & \text{if } \frac{1}{k} \leq R \leq k \\ 1 + (\dfrac{3}{2}\dfrac{1}{R-k+1} + 1)(\dfrac{1}{R-k+1} - 1)^3 & \text{if } R \geq k. \end{cases}$$

Our choices. In steady computations, the Gamma limiter will be chosen. For unsteady computations, and in particular in combination with the low-dissipation superconvergent scheme, the three-entry limiter will be applied.

1.2. Viscous compressible flows

1.2.1. *Model for laminar flows*

1.2.1.1. *Continuous state system*

The compressible Navier–Stokes equations for mass, momentum and energy conservation read

$$\begin{cases} \dfrac{\partial \rho}{\partial t} + \nabla \cdot (\rho \mathbf{u}) = 0, \\ \dfrac{\partial (\rho \mathbf{u})}{\partial t} + \nabla \cdot (\rho \mathbf{u} \otimes \mathbf{u}) + \nabla p = \nabla \cdot \mathcal{T}, \\ \dfrac{\partial (\rho E)}{\partial t} + \nabla \cdot ((\rho E + p)\mathbf{u}) = \nabla \cdot (\mathcal{T} \cdot \mathbf{u}) + \nabla \cdot (\lambda \nabla T), \end{cases} \quad [1.48]$$

where ρ denotes the density ($\mathrm{kg/m^3}$), $\mathbf{u}$ is the velocity ($\mathrm{m/s}$), E is the total energy per mass ($\mathrm{m^2.s^{-2}}$), p is the pressure ($\mathrm{N/m^2}$), given by [1.3], T is the temperature (K) such that $\rho C_v T = E - \frac{1}{2}\rho(u^2 + v^2 + w^2)$ (C_v being the specific heat at constant volume), μ is the laminar dynamic viscosity ($\mathrm{kg/(m.s)}$) and λ is the laminar conductivity. $\mathcal{T}$ is the laminar stress tensor:

$$\mathcal{T} = \mu \left[(\nabla \mathbf{u} + \nabla \mathbf{u}^{\mathrm{T}}) - \frac{2}{3} \nabla . \mathbf{u} \, I \right],$$

where (in 3D) $\mathbf{u} = (u, v, w)$ and

$$\nabla . \mathbf{u} I = \begin{pmatrix} u_x + v_y + w_z & 0 & 0 \\ 0 & u_x + v_y + w_z & 0 \\ 0 & 0 & u_x + v_y + w_z \end{pmatrix},$$

where $u_x = \frac{\partial u}{\partial x}$, $u_y = \frac{\partial u}{\partial y}$, $u_z = \frac{\partial u}{\partial z}$ (idem for v and w). The variation of non-dimensionalized laminar dynamic viscosity and conductivity coefficients μ and λ as functions of the dimensional temperature T are defined by Sutherland's law:

$$\mu = \mu_\infty \left(\frac{T}{T_\infty}\right)^{\frac{3}{2}} \left(\frac{T_\infty + \mathrm{Su}}{T + \mathrm{Su}}\right) \quad \text{and} \quad \lambda = \lambda_\infty \left(\frac{T}{T_\infty}\right)^{\frac{3}{2}} \left(\frac{T_\infty + \mathrm{Su}}{T + \mathrm{Su}}\right),$$

where $\mathrm{Su} = 110°K$ is the Sutherland temperature and the index ∞ denotes reference quantities. The relation linking μ and λ is expressed from the Prandtl laminar number:

$$\mathrm{Pr} = \frac{\mu C_p}{\lambda} \quad \text{with} \quad \mathrm{Pr} = 0.72 \quad \text{for (dry) air},$$

where C_p is the specific heat at constant pressure. In some case, we shall be interested by the steady compressible Navier–Stokes system for a perfect gas, which can be written under a compact form as

$$\nabla \cdot \mathcal{F}^E(W) - \nabla \cdot \mathcal{F}^V(W) = 0 \ \text{ in } \Omega \subset \mathbb{R}^3, \tag{1.49}$$

where $W = {}^t(\rho, \rho\mathbf{u}, \rho E)$ is the conservative flow variables vector and vector $\mathcal{F}^E$ represents the Euler fluxes:

$$\mathcal{F}^E(W) = {}^t\left(\rho\mathbf{u},\ \rho u_1\mathbf{u} + p\mathbf{e}_1,\ \rho u_2\mathbf{u} + p\mathbf{e}_2,\ \rho u_3\mathbf{u} + p\mathbf{e}_3,\ \rho\mathbf{u}H\right),$$

where $H = E + p/\rho$ is the total enthalpy. We write in short the viscous fluxes and viscous stress tensor σ as follows:

$$\mathcal{F}^V(W) = [0, \sigma, -(\mathbf{q} - \mathbf{u} \cdot \sigma)]^T \quad ; \quad \sigma = \mu(\nabla\mathbf{u} + \nabla\mathbf{u}^{\mathbf{T}}) - \frac{2}{3}\mu\nabla \cdot \mathbf{u}\mathbf{I},$$

with μ representing the constant viscosity. The heat flux $\mathbf{q}$ is given by Fourier's law $\mathbf{q} = -\lambda\nabla T$, where λ is the heat conduction. With the standpoint of mesh adaptation, we consider for theoretical developments regular C^3 solution fields. The five unknowns are in functional product space:

$$\mathcal{V} = \left[H^1(\Omega) \cap C^3(\bar{\Omega})\right] \times \left[H_0^1(\Omega) \cap C^3(\bar{\Omega})\right]^3 \times \left[H^1(\Omega) \cap C^3(\bar{\Omega})\right], \quad [1.50]$$

assuming adiabatic conditions on walls. We formulate the Navier–Stokes model in a compact variational formulation:

Find $W \in \mathcal{V}$ such that $\forall \psi \in \mathcal{V}$, $(\Psi(W), \psi) = 0$ with $\Psi = \Psi^E + \Psi^\Gamma + \Psi^V$, [1.51]

where $\psi = (\psi_\rho, \psi_{\rho u}, \psi_{\rho v}, \psi_{\rho w}, \psi_{\rho E})^T$. The Euler term Ψ^E relies on the usual Euler fluxes $\mathcal{F}^E$ denoted $\mathcal{F}$ in [1.1]:

$$(\Psi^E(W), \psi) = \int_\Omega \psi \cdot \nabla \cdot \mathcal{F}^E(W)\, \mathrm{d}\Omega. \quad [1.52]$$

Term Ψ^Γ holds for boundary fluxes contribution, which we denote shortly (see equation [1.4]):

$$(\Psi^\Gamma(W), \psi) = \int_\Gamma \psi \cdot \hat{\mathcal{F}}(W) \cdot \mathbf{n}\, \mathrm{d}\Gamma.$$

Viscous fluxes $\mathcal{F}^V$ provide seven terms:

$$(\Psi^V(W), \psi) = \int_\Omega \psi \cdot \nabla \cdot \mathcal{F}^V(W)\, \mathrm{d}\Omega = \sum_{k=1}^{7} T_k^V. \quad [1.53]$$

The first three terms come from moment equations and depend only on $\psi_{\rho\mathbf{u}} = (\psi_{\rho u}, \psi_{\rho v}, \psi_{\rho w})^T$:

$$T_1^V = \int_\Omega \psi_{\rho\mathbf{u}} \cdot \nabla \cdot (\mu \nabla \mathbf{u})\, \mathrm{d}\Omega$$
$$T_2^V = \int_\Omega \psi_{\rho\mathbf{u}} \cdot \nabla \cdot \left(\mu (\nabla \mathbf{u})^T\right) \mathrm{d}\Omega$$
$$T_3^V = -\frac{2}{3} \int_\Omega \psi_{\rho\mathbf{u}} \cdot \nabla \cdot (\mu\, (\nabla \cdot \mathbf{u})\, \mathbf{I})\, \mathrm{d}\Omega.$$

The last four terms are derived from the energy equation:

$$
\begin{aligned}
T_4^V &= \int_\Omega \psi_{\rho E} \nabla \cdot (\lambda \nabla T) \, \mathrm{d}\Omega \\
T_5^V &= \int_\Omega \psi_{\rho E} \nabla \cdot \left(\mu \, \mathbf{u} \cdot (\nabla \mathbf{u})^T\right) \mathrm{d}\Omega \\
T_6^V &= \int_\Omega \psi_{\rho E} \nabla \cdot (\mu \, \mathbf{u} \cdot \nabla \mathbf{u}) \, \mathrm{d}\Omega \\
T_7^V &= -\frac{2}{3} \int_\Omega \psi_{\rho E} \nabla \cdot (\mu \, \mathbf{u} \cdot ((\nabla \cdot \mathbf{u}) \, \mathbf{I})) \, \mathrm{d}\Omega.
\end{aligned}
$$

1.2.1.2. *Variational discrete formulation*

For the spatial semi-discrete model, we consider the extension of the mixed finite element-finite volume (previous section) to viscous case. We can formulate it under the form of a finite element variational formulation. We assume that Ω is covered by a finite element mesh $\mathcal{H}$ composed of simplicial elements, denoted as K. Let us introduce the following approximation space:

$$
\mathcal{V}_h = \left\{ \psi_h \in \mathcal{V} \mid \psi_{h|K} \text{ is affine } \forall K \in \mathcal{H} \right\}, \qquad [1.54]
$$

where $\mathcal{V}$ is defined by [1.50]. The weak discrete formulation is written as follows:

Find $W_h \in \mathcal{V}_h$ such that $\forall \psi_h \in \mathcal{V}_h, \ (\Psi_h(W_h), \psi_h) = 0,$

with

$$
\begin{aligned}
(\Psi_h(W_h), \psi_h) = &\int_\Omega \psi_h \cdot \nabla \cdot \mathcal{F}_h^E(W_h) \, \mathrm{d}\Omega + \int_\Gamma \psi_h \cdot \hat{\mathcal{F}}_h(W_h).\mathbf{n} \, \mathrm{d}\Gamma \qquad [1.55] \\
&+ \int_\Omega \psi_h \cdot \nabla \cdot \mathcal{F}_h^V(W_h) \, \mathrm{d}\Omega + \int_\Omega \psi_h \cdot D_h(W_h) \, \mathrm{d}\Omega. \qquad [1.56]
\end{aligned}
$$

The fourth RHS term in the previous formulation is an added numerical diffusion dedicated to numerical stability. Indeed, the D_h term holds for the difference between the Galerkin central-differences approximation and the second-order

Godunov approximation. For smooth fields, the D_h term is a third-order term with respect to the mesh size parameter h.

Taking in system [1.55] the P^1 interpolation of the fluxes $\mathcal{F}$ as discretization principle produces a finite-element scheme, which is identical to the central-differenced finite-volume scheme built on the median dual cells, thus

$$\forall W \in \mathcal{V}_h \cup \mathcal{V}\,,\ \mathcal{F}_h^E(W) = \Pi_h \mathcal{F}^E(W) \text{ and } \hat{\mathcal{F}}_h(W) = \Pi_h \hat{\mathcal{F}}(W). \quad [1.57]$$

However, for the viscous term, we cannot move the interpolation outside the flux because viscous fluxes are expressed in terms of primitive variables. Let f be the transformation function from primitive variables $\tilde{U} = (\rho, \mathbf{u}, T)$ into conservatives ones $W = (\rho, \rho\mathbf{u}, \rho E)$, we set

$$\mathcal{F}_h^V(W) = \mathcal{F}_h^V(f(\tilde{U})) = \mathcal{F}^V(f(\Pi_h \tilde{U})) = \mathcal{F}^V(f(\Pi_h f^{-1}(W))).$$

In other words, our discretization consists of P^1 interpolating the primitive variables in the above viscous terms $\{T_i^V\}_{i=1..7}$ for our discretization principle. This completes the definition of the discrete system under study.

1.2.2. *Boundary conditions spatial discretization*

Boundary conditions are imposed element-wise, for example, a boundary flux is computed for each boundary edge/face and added to each of its vertices.

1.2.3. *No-slip boundary condition*

For no-slip boundary conditions, $\mathbf{u} = 0$ is strongly enforced at each iteration. Consistently, we impose $\Phi_\rho = 0$ at the boundary. The energy flux is fixed according to the desired temperature behavior: for an adiabatic wall, it is null, and for an isothermal wall, the energy variable is enforced similarly to the velocity.

1.2.4. *Slip boundary condition*

For this boundary condition, we impose weakly

$$\mathbf{u} \cdot \mathbf{n} = 0. \quad [1.58]$$

To this end, we compute the flux Φ between the state on the boundary W and a mirror state $\overline{W}$:

$$\Phi_{Slip} = \Phi^{HLLC}(W, \overline{W}, \mathbf{n}),$$

where

$$W = \begin{pmatrix} \rho \\ \rho\mathbf{u} \\ \rho E \end{pmatrix} \quad \text{and} \quad \overline{W} = \begin{pmatrix} \rho \\ \rho\mathbf{u} - 2\,\rho\,(\mathbf{u}\cdot\mathbf{n})\,\mathbf{n} \\ \rho E \end{pmatrix}.$$

If condition [1.58] is satisfied, then $W = \overline{W}$ and thus $\Phi^{HLLC}(W, \overline{W}, \mathbf{n}) = F(W)\cdot\mathbf{n}$. Moreover, if W verifies relation [1.58], then $F(W)\cdot\mathbf{n}$ simplifies to

$$\Phi_{Slip} = F(W)\cdot\mathbf{n} = (0, p\,\mathbf{n}, 0)^t.$$

Therefore, if the desired condition is satisfied, then the boundary flux reduces to its well-known commonly used form. Nevertheless, the state W on the boundary does not satisfy this condition unless it is imposed strongly, which is not possible as we will no more conserve the mass. This problem is solved by computing the flux between the state and its mirror state. This flux depends on the considered approximate Riemann solver.

1.2.5. *Influence stencil*

As boundary conditions fluxes are computed element-wise, they involve at most a dependency between a vertex and its neighbors. The total contribution of boundary terms to the flux of a given vertex can be summed up as

$$\mathbf{\Gamma}_i = \sum_{F^{bdy} \ni P_i} \Phi^{bdy}_{i,F^{bdy}}(W_i, W_j, W_k), \qquad [1.59]$$

where F^{bdy} are the boundary elements containing the vertex P_i.

1.2.6. *Spalart–Allmaras one equation turbulence model*

Various forms of the Spalart–Allmaras (SA) turbulence model exist. The original Spalart–Allmaras one equation turbulence model is written as follows (Spalart and Allmaras 1992):

$$\frac{\partial \tilde{\nu}}{\partial t} + \mathbf{u} \cdot \nabla \tilde{\nu} = c_{b1}[1 - f_{t2}]\tilde{S}\tilde{\nu} - \left[c_{w1} f_w - \frac{c_{b1}}{\kappa^2} f_{t2}\right] \left(\frac{\tilde{\nu}}{d}\right)^2 + \frac{1}{\sigma}\left[\nabla \cdot ((\nu + \tilde{\nu})\nabla \tilde{\nu}) + c_{b2}\|\nabla \tilde{\nu}\|^2\right] + f_{t1}\Delta \mathbf{u}^2,$$

where $\tilde{\nu}$ is the kinematic eddy turbulent viscosity. Hereafter, we detail some of these variants.

1.2.7. *SA one-equation model without trip and without f_{t2} term*

In the standard SA model, the trip term is being left out, that is, $f_{t1} = 0$. Moreover, some implementations ignore also the f_{t2} term as it is argued that if the trip is not included, then f_{t2} is not necessary (Eça et al. 2007). In this monograph, this simplified version is considered. It reads in pseudo-vector notations:

$$\frac{\partial \tilde{\nu}}{\partial t} + u_j \frac{\partial \tilde{\nu}}{\partial x_j} = c_{b1}\tilde{S}\tilde{\nu} - c_{w1} f_w \left(\frac{\tilde{\nu}}{d}\right)^2 + \frac{1}{\sigma}\left[\frac{\partial}{\partial x_j}\left((\nu + \tilde{\nu})\frac{\partial \tilde{\nu}}{\partial x_j}\right) + c_{b2}\frac{\partial \tilde{\nu}}{\partial x_i}\frac{\partial \tilde{\nu}}{\partial x_i}\right].$$

However, we prefer to write it under the following form, which is more appropriate for its discretization with the mixed element-volume method:

$$\frac{\partial \rho\tilde{\nu}}{\partial t} + \frac{\partial u_j \rho\tilde{\nu}}{\partial x_j} = \rho c_{b1}\tilde{S}\tilde{\nu} - \rho c_{w1} f_w \left(\frac{\tilde{\nu}}{d}\right)^2 + \frac{\rho}{\sigma}\left[\frac{\partial}{\partial x_j}\left((\nu + \tilde{\nu})\frac{\partial \tilde{\nu}}{\partial x_j}\right) + c_{b2}\frac{\partial \tilde{\nu}}{\partial x_i}\frac{\partial \tilde{\nu}}{\partial x_i}\right].$$

REMARK 1.5.– This is not a conservative model. If conservative form of the Spalart–Allmaras is foreseen, we have to consider the variation of Catris and Aupoix (2000). □

The turbulent eddy viscosity is computed from

$$\mu_t = \rho\tilde{\nu} f_{v1},$$

where

$$f_{v1} = \frac{\chi^3}{\chi^3 + c_{v1}^3} \quad \text{and} \quad \chi = \frac{\tilde{\nu}}{\nu} \quad \text{with} \quad \nu = \frac{\mu}{\rho}.$$

Additional definitions are given by the following equations:

$$\tilde{S} = \Omega + \frac{\tilde{\nu}}{\kappa^2 d^2} f_{v2} \quad \text{where} \quad \Omega = \|\nabla \times \mathbf{u}\|.$$

The magnitude of the vorticity is computed from the vorticity tensor where each component is given by $\omega_{ij} = \frac{1}{2}\left(\frac{\partial u_i}{\partial x_j} - \frac{\partial u_j}{\partial x_i}\right)$ and $\Omega = \sqrt{2 \sum_{i,j=1..3} \omega_{ij}\omega_{ij}}$. Symbol d holds for the distance from the field point to the nearest wall and

$$f_{v2} = 1 - \frac{\chi}{1 + \chi f_{v1}}.$$

Note that we have the following relations:

$$\chi f_{v1} = \frac{\tilde{\nu} f_{v1}}{\nu} = \frac{\nu_t}{\nu} \quad \Longrightarrow \quad 1 - \frac{\chi}{1 + \chi f_{v1}} = 1 - \frac{\tilde{\nu}}{\nu + \nu_t}.$$

The constants are

$$\sigma = \frac{2}{3} \qquad c_{b1} = 0.1355 \;\; c_{b2} = 0.622 \;\; \kappa = 0.41$$

$$c_{w1} = \frac{c_{b1}}{\kappa} + \frac{1 + c_{b2}}{\sigma} \qquad c_{w2} = 0.3 \qquad c_{w3} = 2 \qquad c_{v1} = 7.1.$$

Finally, the function f_w is computed as follows:

$$f_w = g\left(\frac{1 + c_{w3}^6}{g^6 + c_{w3}^6}\right)^{1/6} \quad \text{with } g = r + c_{w2}\left(r^6 - r\right) \text{ and } r = \min\left(\frac{\tilde{\nu}}{\tilde{S}\kappa^2 d^2}, 10\right).$$

According to Spalart and Rumsey (2007), for the boundary conditions the following values are chosen:

$$\tilde{\nu}_{wall} = 0 \qquad \tilde{\nu}_{farfield} \in [3\nu_\infty, 5\nu_\infty].$$

1.2.8. *"Standard" SA one-equation model (without trip)*

It reads in pseudo-vector notations:

$$\frac{\partial \tilde{\nu}}{\partial t} + u_j \frac{\partial \tilde{\nu}}{\partial x_j} = c_{b1}[1 - f_{t2}]\tilde{S}\tilde{\nu} - \left[c_{w1} f_w - \frac{c_{b1}}{\kappa^2} f_{t2}\right] \left(\frac{\tilde{\nu}}{d}\right)^2$$
$$+ \frac{1}{\sigma} \left[\frac{\partial}{\partial x_j}\left((\nu + \tilde{\nu})\frac{\partial \tilde{\nu}}{\partial x_j}\right) + c_{b2} \frac{\partial \tilde{\nu}}{\partial x_i} \frac{\partial \tilde{\nu}}{\partial x_i}\right],$$

where $f_{t2} = c_{t3} \exp\left(-c_{t4}\chi^4\right)$ with $c_{t3} = 1.2$ and $c_{t4} = 0.5$.

1.2.9. *"Full" SA one-equation model (with trip)*

For the full SA model, we have to compute the f_{t1} term:

$$f_{t1} = c_{t1} g_t \exp\left(-c_{t2} \frac{\omega_t^2}{\delta \mathbf{u}_t^2}(d^2 + g_t^2 d_t^2)\right) \quad \text{with} \quad g_t = \min\left(0.1, \frac{\delta \mathbf{u}_t}{\omega_t \delta x_t}\right),$$

where $\delta \mathbf{u}_t$ is the difference between the velocity at the field point and that at the trip (on the wall), δx_t is the grid spacing along the wall at the trip, ω_t is the wall vorticity at the trip and d_t is the distance from the field point to the trip and

$$c_{t1} = 1 \quad \text{and} \quad c_{t2} = 2.$$

The farfield boundary condition is given as

$$0 \leq \tilde{\nu}_{farfield} \leq 1/10\nu_\infty.$$

1.2.10. *Mixed element-volume discretization of SA*

To discretize the turbulent model, the mixed element-volume method is considered. The Euler explicit discretization of the above equation in time and in space leads to

$$\frac{V_i}{\Delta t} \delta \rho \tilde{\nu}_i = -\mathbf{F}_i^n + \mathbf{S}_i^n + \mathbf{Q}_i^n,$$

where $\mathbf{F}$, $\mathbf{S}$ and $\mathbf{Q}$ are, respectively, the Euler, viscous and source flux terms, $\delta \rho \tilde{\nu}_i = \rho_i^{n+1} \tilde{\nu}_i^{n+1} - \rho_i^n \tilde{\nu}_i^n$ and $V_i = |C_i|$.

1.2.10.1. *Convective terms*

First, we describe how the convective part is discretized. The flux on the turbulent variable is given by

$$\mathbf{F}_i^n = \sum_{edge_{ij}} \mathbf{\Phi}_{ij}^{\rho\tilde{\nu}}(\mathbf{W}_i, \mathbf{W}_j, \nu_{ij}),$$

where $\mathbf{W}$ is the conservative vector variable $(\rho, \rho\mathbf{u}, \rho E, \rho\tilde{\nu})$, and ν_{ij} is the edge cell interface non-normalized normals from $\mathbf{p}_i$ to $\mathbf{p}_j$. Here, we have two choices.

1.2.10.2. *Linear convection*

The simplest one is to consider a linear advection of $\rho\tilde{\nu}$:

$$\mathbf{\Phi}_{ij}^{\rho\tilde{\nu}}(\mathbf{W}_i, \mathbf{W}_j, \nu_{ij}) = \begin{cases} \eta\ \rho\tilde{\nu}_i & \text{if} \quad \eta > 0 \\ \eta\ \rho\tilde{\nu}_j & \text{otherwise,} \end{cases} \qquad [1.60]$$

where $\eta = \dfrac{1}{2}\,(\mathbf{u}_i \cdot \nu_{ij} + \mathbf{u}_j \cdot \nu_{ij})$.

1.2.10.3. *Nonlinear convection*

The second choice is nonlinear and has been proposed in Larrouturou (1989). It has been proven that this scheme preserves the maximum principle for the convected turbulent variable. It reads

$$\mathbf{\Phi}_{ij}^{\rho\tilde{\nu}}(\mathbf{W}_i, \mathbf{W}_j, \nu_{ij}) = \mathbf{\Phi}_{ij}^{\rho}(\mathbf{W}_i, \mathbf{W}_j, \nu_{ij}) \cdot \begin{cases} \tilde{\nu}_i & \text{if} \quad \mathbf{\Phi}_{ij}^{\rho}(\mathbf{W}_i, \mathbf{W}_j, \nu_{ij}) > 0 \\ \tilde{\nu}_j & \text{otherwise,} \end{cases} \qquad [1.61]$$

where the density flux $\mathbf{\Phi}_{ij}^{\rho}$ is computed with the appropriate Riemann solver.

1.2.10.4. *Viscous terms*

The viscous terms can been discretized in two different ways.

1.2.10.4.1. First-order method

The classical approach consists of considering the diffusion term as a source term (see source terms paragraph for its discretization), that is, it is discretized under the model form. The dissipation term is discretized with the FEM. Then, we have

$$\nabla \cdot S(\mathbf{W}) = \frac{\rho}{\sigma}\left[\nabla \cdot ((\nu + \tilde{\nu})\nabla\tilde{\nu})\right].$$

The viscous fluxes are

$$\mathbf{S}_i^n = \sum_{K\ni i} \int_K S_i(\mathbf{W})|_K \cdot \nabla\Phi_i \, d\mathbf{x} = -\sum_{K\ni i} \int_{\partial C_i \cap K} S_i(\mathbf{W})|_K \cdot \mathbf{n} \, d\sigma$$
$$= -\sum_{\mathbf{p}_j \in B(\mathbf{p}_i)} \int_{\partial C_{ij}} S_i(\mathbf{W}) \cdot \mathbf{n}_{ij} \, d\sigma, \qquad [1.62]$$

where Φ_i is the P_1 finite element basis function associated with vertex i $B(\mathbf{p}_i)$ is the set of neighbors of i. The effective computation of the integral then reduces to the computations of integrals of the type

$$\int_K \nabla\Phi_i \, \nabla\Phi_j \, d\mathbf{x} \; = \; |K| \, \nabla\Phi_i|_K \, \nabla\Phi_j|_K.$$

We set

$$\tilde{\nu}|_K = \frac{1}{4} \sum_{\mathbf{p}_j \in K} \tilde{\nu}_j \qquad \text{and} \qquad \nabla\tilde{\nu}|_K = \sum_{\mathbf{p}_j \in K} \tilde{\nu}_j \nabla\Phi_j,$$

then the dissipation term $\nabla \cdot \left(\frac{\rho_i}{\sigma}(\nu + \tilde{\nu})\nabla\tilde{\nu} \right)$ is written as

$$\left(S_i^{\text{diss}}\right)|_K = |K| \, \frac{1}{\sigma}\rho_i \left((\nu|_K + \tilde{\nu}|_K) \, \nabla\tilde{\nu}|_K \cdot \nabla\Phi_i|_K \right).$$

But, in that case, *the discretization is only first-order spatially accurate*. The second order of the method is lost.

1.2.10.5. *Second-order method*

To recover the second order of the method, diffusive and dissipative terms are re-written under the following form:

$$\nabla \cdot S(\mathbf{W}) \; = \; \frac{\rho}{\sigma}\left[\nabla \cdot ((\nu + \tilde{\nu})\nabla\tilde{\nu}) + c_{b2}\nabla\tilde{\nu} \cdot \nabla\tilde{\nu}\right]$$
$$= \; \frac{\rho}{\sigma}\left[\nabla \cdot ((\nu + (1 + c_{b2})\tilde{\nu})\nabla\tilde{\nu}) - c_{b2} \, \tilde{\nu} \, \nabla \cdot \nabla\tilde{\nu}\right]$$

using relation $\nabla \cdot (\tilde{\nu} \nabla \tilde{\nu}) = \nabla \tilde{\nu} \cdot \nabla \tilde{\nu} + \tilde{\nu}\, \nabla \cdot \nabla \tilde{\nu}$. Then, the diffusion and dissipation terms are discretized with the FEM using the viscous flux given by relation [1.62]. The diffusion term $\frac{c_{b2}}{\sigma}\, \nabla \cdot (\rho_i \tilde{\nu}_i\, \nabla \tilde{\nu})$ reads

$$\left(S_i^{\mathrm{diff}}\right)|_K = |K|\, \frac{c_{b2}}{\sigma} \rho_i \tilde{\nu}_i \left(\nabla \tilde{\nu}|_K \cdot \nabla \Phi_i|_K\right)$$

and the dissipation term $\nabla \cdot \left(\frac{\rho_i}{\sigma}(\nu + (1 + c_{b2})\tilde{\nu})\nabla \tilde{\nu}\right)$ is given as

$$\left(S_i^{\mathrm{diss}}\right)|_K = |K|\, \frac{\rho_i}{\sigma} \Big((\nu|_K + (1 + c_{b_2})\tilde{\nu}|_K)\; \nabla \tilde{\nu}|_K \cdot \nabla \Phi_i|_K \Big).$$

1.2.10.6. *Source terms*

Finally, the source terms (diffusion, production and destruction for the order one method and, production and destruction for the order two method) are discretized by simple integration on each vertex cell:

$$\text{First order:} \quad \mathbf{Q}_i^n = |C_i| \left(\rho_i c_{b1} \tilde{S}_i \tilde{\nu}_i - \rho_i c_{w1} f_w \left(\frac{\tilde{\nu}_i}{d_i}\right)^2 + \frac{c_{b2}}{\sigma} \rho_i \|\nabla \tilde{\nu}_i\| \right),$$

$$\text{Second order:} \quad \mathbf{Q}_i^n = |C_i| \left(\rho_i c_{b1} \tilde{S}_i \tilde{\nu}_i - \rho_i c_{w1} f_w \left(\frac{\tilde{\nu}_i}{d_i}\right)^2 \right),$$

where $|C_i|$ is the volume of the vertex cell and all variables are point-wise:

$$\tilde{S}_i = \Omega_i + \frac{\tilde{\nu}_i}{\kappa^2 d_i^2} f_{v2}.$$

Influence stencil. As each element produces a viscous flux for each of its vertices, we can deduce the total viscous fluxes for a given cell/vertex i as

$$\mathbf{S}_i = \sum_{K \ni P_i} \Phi_{i,K}^{visc}(W_i, W_j, W_k, W_l), \qquad [1.63]$$

so that the viscous fluxes of a vertex only depend on the ball $\mathcal{V}^1(P_i)$ of first order neighbors of P_i.

1.2.11. *Implicit time integration*

Once the equations have been discretized in space, a set of ordinary differential equations in time is obtained. For an implicit time integration, the semi-discretized RANS system becomes

$$\frac{|C_i|}{\delta t_i^n}\delta W_i = -\mathbf{F}_i^{n+1} + \mathbf{S}_i^{n+1} + \mathbf{Q}_i^{n+1} + \mathbf{\Gamma}_i^{n+1}, \qquad [1.64]$$

where $\delta W_i = W_i^{n+1} - W_i^n$, $|C_i|$ is the area/volume of the finite volume cell associated with vertex i and δt_i^n is the considered time step. After linearization of the RHS, this becomes:

$$\begin{aligned}&\left(\frac{|C_i|}{\delta t_i^n}I_d + \frac{\partial \mathbf{F}_i^n}{\partial W_i} - \frac{\partial \mathbf{S}_i^n}{\partial W_i} - \frac{\partial \mathbf{Q}_i^n}{\partial W_i} - \frac{\partial \mathbf{\Gamma}_i^n}{\partial W_i}\right)\delta W_i\\ &+ \sum_{j\in\mathcal{V}^1(i)}\left(\frac{\partial \mathbf{F}_i^n}{\partial W_j} - \frac{\partial \mathbf{S}_i^n}{\partial W_j} - \frac{\partial \mathbf{Q}_i^n}{\partial W_j} - \frac{\partial \mathbf{\Gamma}_i^n}{\partial W_j}\right)\delta W_j\\ &= -\mathbf{F}_i^n + \mathbf{S}_i^n + \mathbf{Q}_i^n + \mathbf{\Gamma}_i^n,\end{aligned}$$

where $P_j \in \mathcal{V}^1(i)$ is the set of vertices connected to vertex i by an edge. But, the computation of the Jacobian of the second-order inviscid flux $\dfrac{\partial \mathbf{F}_i^n}{\partial W_j}$ is complex and costly (as it involved the second order ball of vertex i, as seen above, while the other terms only involve the first order ball). It is thus approximated by the Jacobian of the first-order inviscid flux $\dfrac{\partial \widetilde{\mathbf{F}}_i^n}{\partial W_j}$, which only depends on the first-order ball (the other Jacobian terms are kept unchanged):

$$\begin{aligned}&\left(\frac{|C_i|}{\delta t_i^n}I_d + \frac{\partial \widetilde{\mathbf{F}}_i^n}{\partial W_i} - \frac{\partial \mathbf{S}_i^n}{\partial W_i} - \frac{\partial \mathbf{Q}_i^n}{\partial W_i} - \frac{\partial \mathbf{\Gamma}_i^n}{\partial W_i}\right)\delta W_i\\ &+ \sum_{j\in\mathcal{V}^1(i)}\left(\frac{\partial \widetilde{\mathbf{F}}_i^n}{\partial W_j} - \frac{\partial \mathbf{S}_i^n}{\partial W_j} - \frac{\partial \mathbf{Q}_i^n}{\partial W_j} - \frac{\partial \mathbf{\Gamma}_i^n}{\partial W_j}\right)\delta W_j\\ &= -\mathbf{F}_i^n + \mathbf{S}_i^n + \mathbf{Q}_i^n + \mathbf{\Gamma}_i^n.\end{aligned}$$

As the RHS still consider the second-order inviscid flux, this acts as an approximated Jacobian but is independent of the residual and thus does not affect the spatial order of the scheme. The first term of the LHS contributes to the diagonal of the matrix and the second term of the LHS (i.e. the sum) contributes to extra-diagonal terms on line i of the matrix. This linearized system can be written in vector form:

$$\mathbf{A}^n\,\delta W^n = \mathbf{R}^n \quad \text{with} \quad \mathbf{A}^n = \frac{|C|}{\delta t^n}\mathbf{I} - \frac{\partial \widetilde{\mathbf{R}}^n}{\partial \mathbf{W}} \quad \text{and} \quad \delta \mathbf{W}^n = \mathbf{W}^{n+1} - \mathbf{W}^n,$$

where $\mathbf{R}^n = -\mathbf{F}^n + \mathbf{S}^n + \mathbf{Q}^n + \mathbf{\Gamma}^n$ and $\widetilde{\mathbf{R}}^n = -\widetilde{\mathbf{F}}^n + \mathbf{S}^n + \mathbf{Q}^n + \mathbf{\Gamma}^n$.

This linear system is solved at each flow solver iteration. To solve it, we follow the SGS approach based on lower-upper symmetric Gauss-Seidel (LU-SGS) implicit solver initially introduced by Jameson and Yoon (1987) and fully developed by Luo et al. (1998, 2001), Sharov and Nakahashi (1997) and Sharov et al. (2000). The SGS is very attractive because it uses an edge-based data structure, which can be efficiently parallelized with p-threads (Alauzet and Loseille 2009; Sharov et al. 2000).

The INRIA code `Wolf` involves all the above features, in short the Spalart–Allmaras model, MUSCL, Piperno and Gamma limiters and LU-SGS implicit formulation.

1.3. A multi-fluid incompressible model

1.3.1. *Introduction*

Flows with interfaces have been rather early addressed in the history of computer aided-simulation (see, for example, Harlow and Welch 1965). We give a first example of continuous CFD model for the flow of two non-miscible incompressible flows, using a level set representation of the interface. A discretization based on P^1 interpolation is then defined. Some idea about the resulting accuracy is given.

1.3.2. *Bi-fluid incompressible Navier–Stokes equations*

Let us consider the model problem of two incompressible immiscible fluids moving in a closed vessel Ω under the influence of gravity and without interface tension. Then, the bi-fluid incompressible Navier–Stokes system reads

$$\rho\frac{\partial \boldsymbol{U}}{\partial t} + \rho\nabla\cdot(\boldsymbol{U}\otimes\boldsymbol{U}) = \nabla\cdot(\nu(\rho)\nabla\boldsymbol{U}) - \nabla p + \rho\boldsymbol{g} \text{ in } \Omega \qquad [1.65]$$

$$\nabla\cdot\boldsymbol{U} = 0 \text{ in } \Omega \qquad [1.66]$$

$$\frac{\partial \rho}{\partial t} + \nabla\cdot(\rho\boldsymbol{U}) = 0 \text{ in } \Omega, \qquad [1.67]$$

where $\boldsymbol{U}$ denotes the fluid velocity, p is the pressure, ρ is the density and $\boldsymbol{g}$ is the gravity volumic force. In this formulation, the density takes only two real positive values ρ_l and ρ_g in two subdomains separated by an *free interface* Γ_f. This model is mathematically well-posed when the viscosity $\nu(\rho)$ is not zero (see for example (Simon 1990)). In this chapter, we only consider a two-fluid inviscid model, that is, $\nu(\rho) = 0$.

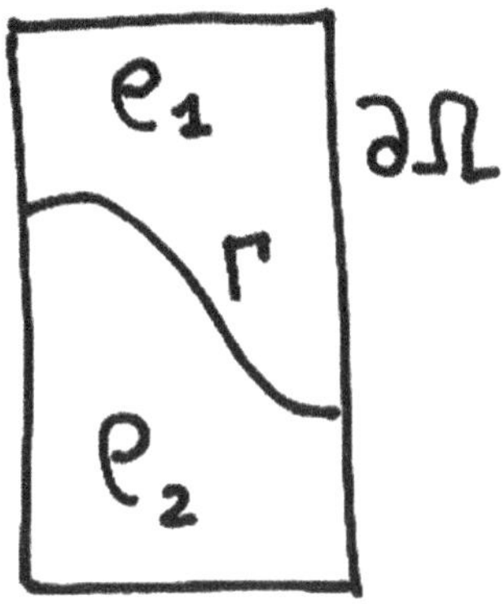

Figure 1.8. *Computational domain*

The interface Γ_f (Figure 1.8) is assumed to be a smooth enough surface, and in particular to have a well-defined normal $\mathbf{n}$:

$$\mathbf{n}\delta_\rho = \frac{1}{\rho_l - \rho_g}\nabla\rho,$$

where δ_ρ denotes the Dirac delta function on the interface Γ_f,

$$(\mathbf{n}\delta_\rho, \mathbf{f})_{(\mathcal{D}')^2\times(\mathcal{D})^2} = \int_{\Gamma_f} \mathbf{f}\cdot\mathbf{n}d\Gamma_f.$$

To fix the ideas, ρ_l is the liquid density and ρ_g is the gas density. This system needs to be completed by *boundary conditions*. We assume a slip condition ($\boldsymbol{U}\cdot\mathbf{n}_{\partial\Omega} = 0$ on $\partial\Omega$ where $\mathbf{n}_{\partial\Omega}$ is the boundary normal) on some part of the boundary and a no-slip condition ($\boldsymbol{U} = 0$) on the other part. It is also possible to prescribe the pressure (we do not go into details).

Let us rewrite the Navier–Stokes system [1.65–1.67] in the *Level Set framework* as in Osher and Sethian (1988) (some very preliminary attempts were presented in

Dervieux and Thomasset 1981). We consider a smooth function ϕ, the level set (iso-value) 0 of which always represents the interface. The interface is then defined by the Heaviside step function H applied to ϕ:

$$H(\phi) = \begin{cases} 0 & \text{if } \phi < 0 \\ 1 & \text{if } \phi \geq 0. \end{cases}$$

The density is constant in each fluid and can be rewritten as a function of ϕ: $\rho(\phi) = \rho_g + (\rho_l - \rho_g)H(\phi)$. Equation [1.67] is then replaced by the interface advection equation

$$\phi_t + \nabla \cdot (\boldsymbol{U}\phi) = 0, \qquad [1.68]$$

and the governing equations for the fluid velocity $\boldsymbol{U}$ and pressure p, that is, relations [1.65] and [1.66] become

$$\rho(\phi)\frac{\partial \boldsymbol{U}}{\partial t} + \rho(\phi)\nabla \cdot (\boldsymbol{U} \otimes \boldsymbol{U}) + \nabla p - \rho(\phi)\boldsymbol{g} = 0 \qquad [1.69]$$

$$\nabla \cdot \boldsymbol{U} = 0. \qquad [1.70]$$

1.3.3. *Finite element approximation*

The system of equations [1.69–1.70] is solved explicitly with a standard P_1-P_1 projection finite element method as in Lesage et al. (2007). The interface is advected by means of equation [1.68], which is solved by a vertex-centered finite volume scheme as mentioned in section 1.1. We first introduce some notations. Let Ω_h be a discretization of the domain Ω of $\mathbb{R}^d$ and $\mathcal{T}_h$ a mesh of Ω_h. The discretized domain Ω_h is assumed to be identical to Ω and thus denoted Ω in the following. For the sake of simplicity, the subscript h associated with the spatial discretization is also omitted for other symbols. $V = \{\psi \in \mathcal{C}^0(\bar{\Omega}) \mid \psi|_T \text{ is affine } \forall T \in \mathcal{T}_h\}$ is the usual P_1 finite element space. V is spanned by the set of basis functions ψ_i, where ψ_i verifies for any vertex $\mathbf{x}_i$ of $\mathcal{T}_h$, $\psi_i(\mathbf{x}_i) = 1$ and $\forall j \neq i$, $\psi_i(\mathbf{x}_j) = 0$. Let $\mathbf{V} = V^d$, where d is the space dimension. The discretized multi-fluid variables are

$$\mathbf{U} = \sum_i \mathbf{U}_i\psi_i, \; p = \sum_i p_i\psi_i \text{ and } \phi = \sum_i \phi_i\psi_i.$$

1.3.3.1. *Transfer operator into* $\mathbf{V}$

For any $u \in \mathbf{L}^2(\Omega)$, we denote by $\mathcal{P}u : \mathbf{L}^2 \mapsto V$ the function such that for any vertex $\mathbf{x}_i$ of $\mathcal{T}_h$,

$$\mathcal{P}u(\mathbf{x}_i) = \frac{\int_\Omega u\psi_i \,\mathrm{d}\mathbf{x}}{\int_\Omega \psi_i \,\mathrm{d}\mathbf{x}}.$$

And, for all $\mathbf{U} = (u, v) \in (\mathbf{L}^2(\Omega))^2$, we denote by $\mathcal{P}\mathbf{U} = (\mathcal{P}u, \mathcal{P}v)$ the transfer into $\mathbf{V}$. The transfer operator $\mathcal{P}$ transforms a discrete field that is constant by element into a discrete field that is continuous and piecewise linear.

The global algorithm for advancing in time follows.

Stage 1 (prediction step). The momentum is explicitly predicted due to equation [1.69] by omitting the pressure term:

$$\bar{\mathbf{U}}_i = \mathbf{U}_i^n - \frac{\Delta t}{|C_i|} \int_\Omega \psi_i \left(\nabla \cdot (\mathbf{U} \otimes \mathbf{U}) - \mathbf{g}\right) \mathrm{d}\mathbf{x},$$

where $|C_i|$ is the vertex cell volume given by the following mass-lumped coefficient:

$$|C_i| = \sum_j \int_\Omega \psi_i \psi_j \,\mathrm{d}\mathbf{x}$$

and where $\mathbf{U}$ in RHS can be taken as $\mathbf{U}^n$ (time-explicit) or $\bar{\mathbf{U}}$ (time-implicit).

Stage 2 (projection step). The projection step evaluates the pressure using the predicted velocity and updates the velocity by imposing the null divergence constraint (relation [1.70]). To this end, the elliptic system

$$\int \frac{1}{\rho} \nabla p^{n+1} \cdot \nabla \psi \,\mathrm{d}\mathbf{x} = \frac{1}{\Delta t} \int \nabla \psi \cdot \bar{\mathbf{U}} \,\mathrm{d}\mathbf{x} \quad \forall\, \psi \in V$$

is solved in V by a finite element method. The correction based on the pressure gradient is constant by element. We transfer it into $\mathbf{V}$ and add to the velocity

$$\mathbf{U}^{n+1} = \bar{\mathbf{U}} + \Delta t\, \mathcal{P}\left(\frac{1}{\rho} \nabla p^{n+1}\right) \quad \text{and} \quad \mathbf{U}^{n+1} = 0 \text{ on } \partial\Omega.$$

Stage 3 (level set advection and re-distancing). Function ϕ is advected by means of equation [1.68] with $\mathbf{U}^n$ from time level n to time level $n+1$. This equation is solved by a second-order vertex-centered finite volume scheme. In a classical manner, the level set function is then re-distanced by replacing it by a distance function $\tilde{\phi}^{n+1}$, which is the signed distance to the set $\{\phi^{n+1} = 0\}$. For re-distancing, several authors recommend building a signed distance by iterating until steady-state a

Hamilton–Jacobi equation (see Sussman et al. 1998; Sethian 1999; Osher and Fedkiw 2002). In the present model, the method of Smolianski (2005) is followed and the adopted option is to rebuild a distance from a geometrical algorithm, defining ϕ_{dist} at each vertex as the minimum distance to discrete zero level of ϕ. At the end of Stage 3, a global volume conservation correction slightly different from that of Smolianski (2005) is applied; a small perturbation C_ϕ is added to ϕ in each point of the domain in order to solve iteratively the following conservation relation:

$$\overline{\phi_h}^{n+1} = \phi_h^{n+1} + C_{\phi_h}, \int H(\overline{\phi_h}^{n+1})dv = \int H(\phi_h^n)dv.$$

1.3.3.2. *Interface thickening*

In equation [1.69], the Heaviside function $H(\phi)$, that defines the density, is replaced by the thickened version H_η defined as follows:

$$H_\eta(\phi) = \begin{cases} 0 & \text{if } \phi < -\eta, \\ \frac{1}{2}(1 + \frac{\phi}{\eta} + \frac{1}{\pi}\sin(\frac{\pi\phi}{\eta})) & \text{if } |\phi| \le \eta, \\ 1 & \text{if } \phi > \eta. \end{cases}$$

In practice, the parameter η is chosen as $\eta = k\ \Delta x$ with $k \approx 3$. This interface thickening reinforces the scheme stability but it carries an extra feature to tune and reduces accuracy to first order.

Once $\tilde{\phi}^{n+1}$ is obtained, we deduce ρ^{n+1}. We refer to Lesage et al. (2007) for further details.

REMARK 1.6.– Usually, a collocated $P_1 - P_1$ approximation of the incompressible Navier–Stokes equations is not stable. The stabilization is obtained by the introduction of the transfer $\mathcal{P}$, which introduces a stabilization of same type as the Rhie–Chow stabilization (Rhie and Chow 1983).

REMARK 1.7.– The above simplified projection stage induces a strong dependency of the solution to time step size, which can be reduced by introducing an auxiliary time step in the projection stage.

1.3.4. *Error estimate for the level set advection*

This section is devoted to justifying the re-distancing and gives elements to choose the spatial discretization. Let us assume that we know how to numerically advect the level set ϕ by approximating it with a discrete level set function ϕ_h. Since ϕ is advected with a first-order hyperbolic model, according to variational theory, the usual convergence property for ϕ_h on non-regular meshes is a convergence in L^p, but we can manage a high-order k of accuracy in that norm. Let us examine the consequence for the corresponding $\chi_\phi = H(\phi)$ functions.

PROPOSITION 1.1.– Let ϕ be a $L^p(Q)$ function where $Q = \Omega\times]0,T[$ is the flow integration domain in space and time. Let $(\phi_h)_h$ a sequence of $L^p(Q)$ and we assume that

$$\begin{array}{l} meas(-\zeta \leq \phi \leq \zeta) \leq K_1\zeta \\ \parallel \phi_h - \phi \parallel_{L^p(Q)} \leq K_2 h^k, \end{array} \qquad [1.71]$$

with h, ζ sufficiently small, k the convergence order on ϕ and K_1,K_2 independent of h, ζ then for all real number $q \geq 1$, there is a constant $C(q)$ independent of h such as[6]

$$\parallel H(\phi_h) - H(\phi) \parallel_{L^q(Q)} \leq C(q) h^{\frac{2k}{3q}}. \square \qquad [1.73]$$

6 PROOF.– The above integral can be analyzed as follows:

$$\int_Q |H(\phi_h) - H(\phi)| d\Omega dt \leq meas(|\phi| \leq \zeta) + meas(|\phi_h| \leq \zeta)$$

$$+meas(\phi \geq \zeta, \phi_h \leq -\zeta) + meas(\phi \leq -\zeta, \phi_h \geq \zeta)$$

$$\leq meas(|\phi| \leq \zeta) + meas(|\phi_h| \leq \zeta) + meas(|\phi_h - \phi| \geq 2\zeta)$$

but $meas(|\phi| \leq \zeta) \leq K_1\zeta$, according to [1.71] and $meas(|\phi_h| \leq \zeta) \leq meas(|\phi| \leq 2\zeta) + meas(|\phi_h - \phi| \geq \zeta)$, that is $meas(|\phi_h| \leq \zeta) \leq 2K_1\zeta + meas(|\phi_h - \phi| \geq \zeta)$ thus $\int_Q |H(\phi_h) - H(\phi)| d\Omega dt \leq 3K_1\zeta + 2meas(|\phi_h - \phi| \geq \zeta)$. The last sum part can be estimated as follows:

$$meas(|\phi_h - \phi| \geq \zeta) = \int_{|\phi_h - \phi|^p \geq \zeta^p} 1\, d\Omega dt \leq \frac{1}{\zeta^p} \int_{|\phi_h - \phi|^p \geq \zeta^p} |\phi_h - \phi|^p d\Omega dt$$

$$\leq \frac{1}{\zeta^p} ||\phi_h - \phi||^p_{L^p(Q)} \qquad [1.72]$$

and with [1.72], we get $\int_Q |H(\phi_h) - H(\phi)| d\Omega dt \leq 3K_1\zeta + \frac{2}{\zeta^p} ||\phi_h - \phi||^p_{L^p(Q)}$. By choosing $\zeta = h^\theta$, $\theta = \frac{pk}{p+1}$, we get

$$\int_Q |H(\phi_h) - H(\phi)| d\Omega dt \leq 3K_1 h^\theta + 2K_2^2 h^{-p\theta} h^{pk} \leq (3K_1 + 2K_2^2) h^{\frac{pk}{p+1}},$$

thus $\int_Q |H(\phi_h) - H(\phi)| d\Omega dt \leq (3K_1 + 2K_2^2) h^{\frac{pk}{p+1}}$ or equivalently, for $q \geq 1$: $||H(\phi_h) - H(\phi)||_{L^q} \leq K_5 h^{\frac{pk}{q(p+1)}}$. $\square$

In the global scheme presented, we can consider that $q = 1, k = p = 2$, which gives a convergence of at least $4/3$, better than order one.

REMARK 1.8.– Estimate [1.73] is not optimal. The order of accuracy is between $pk/(q(p+1))$ and k. However, the possible loss of accuracy shown in [1.73] is related to inaccuracy on the interface when ϕ_h is not sufficiently smooth. In practice, ϕ_h is re-initialized by being set to a signed distance to interface, although that extra step introduces other type of errors. In the sequel, we shall refer to the case $p = 2, q = 1$, which gives an order between $\frac{2k}{3}$ and k. □

REMARK 1.9.– The above analysis easily extends to a thickened interface function H_η, giving an estimate for $||H_\eta(\phi_h) - H_\eta(\phi)||_{L^q}$. Now the exact context should rely on H, and the deviation

$$||H_\eta(\phi) - H(\phi)||_{L^q} \leq K_6\eta$$

needs to be taken into account in the global error estimate. □

REMARK 1.10.– The above analysis suggests that in order to get a global second-order convergence for a level-set multifluid calculation, a second-order Navier–Stokes approximation should be combined with a third-order accurate advection of the function ϕ. □

REMARK 1.11.– The above analysis applies only if a second-order convergence of the velocity is true. The latter convergence is generally not reached when the viscosity is discontinuous across the interface. □

1.3.5. *Provisional conclusion on scheme accuracy*

The proposed level-set $P1$ approximation, because of the level set formulation, is not advecting the interface with first-order accuracy, but the analysis shows that the accuracy can be bounded to order $3/2$, as concerns the interface capture. This takes into account neither any thickening of interface, nor the accuracy of interface representation in moment and pressure equations.

The incompressible diphasic component of the multi-physics platform *NiceFlow* of Lemma[7], which will be used in this book, involves all the above features.

7 Available at: https://www.lemma-ing.com/.

1.4. Appendix: circumcenter cells

Median-based cells (Figure 1.9) are the exact counterbalance of the P^1 finite element formulation. They are well adapted to non-stretched unstructured meshes. It was observed (in particular in Barth 1994) that for certain highly stretched meshes, the median tessellation results in an important deterioration of the accuracy. In that case, according to Barth (1994), it is recommended to replace the median cells by the so-called circumcenter cells. Note, however, that in most of our viscous *mesh adaptative* application, the adequation of the mesh allows keeping the median option.

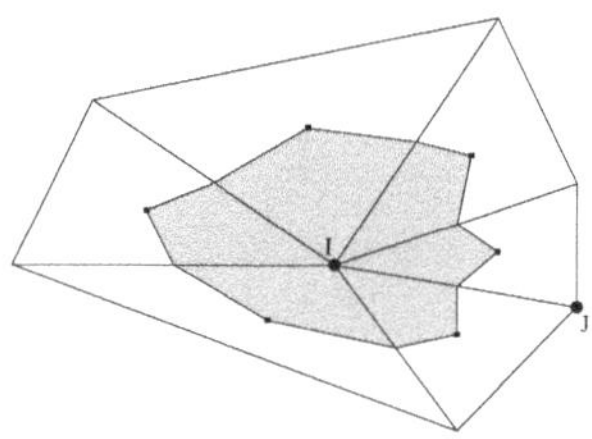

Figure 1.9. *Median cell construction in 2D*

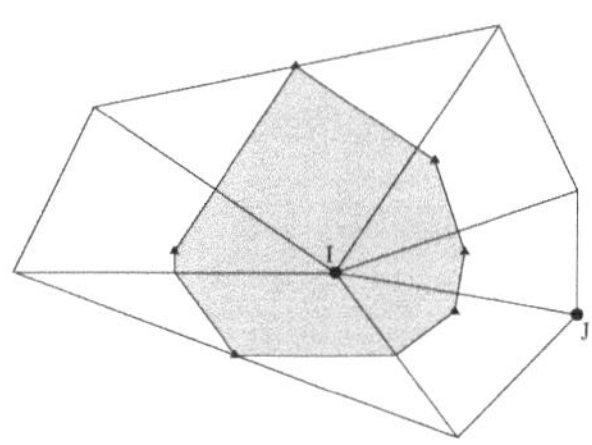

Figure 1.10. *Circumcenter cell construction in 2D*

1.4.1. *Two-dimensional circumcenter cells*

According to an idea of Barth (1994), a cell is built around each vertex by joining the mid-edges with the center of the smallest circle containing the considered triangle. We observe that this center is located at middle of largest edge in a triangle involving an angle larger or equal to 90°. As a result, a mesh made of rectangular triangles has cells, which are rectangles (see Figure 1.12).

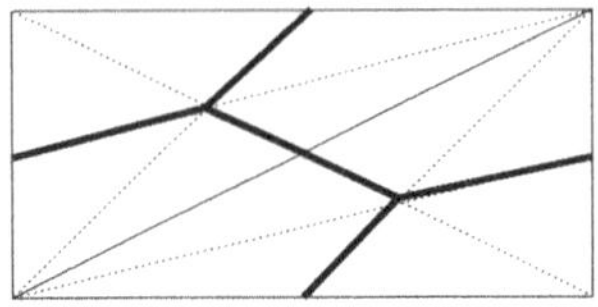

Figure 1.11. *Trace of median division on two elements*

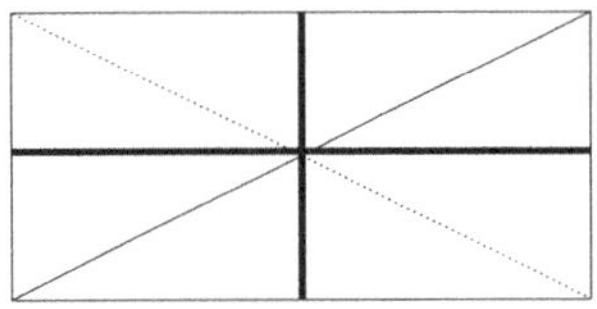

Figure 1.12. *Trace of circumcenter division on the same two elements*

1.4.2. *Three-dimensional circumcenter cells*

A 3D circumcenter cell can be defined as the union of polyhedra such that, for a given element, the common surface between two neighboring cells joins:

– the middle of the edge connecting these two vertices;

– the "surface center" of the faces of the element having this edge in common;

– the "volume center" of the element;

where

– the *"surface center"* of a given face is the center of its circumscript circle if it comprises only acute angles, otherwise, it is the middle of its longest edge;

– the *"volume center"* of an element is the center of its circumscript sphere if this center is located inside the element; otherwise, it is the *"surface center"* of the face which is closest to the center of its circumscript sphere.

In practical applications, we use the circumcenter cells for stretched meshes with aspect ratios larger than 10 and for Cartesian meshes. Indeed, it is proposed in Gourvitch et al. (2004) to combine the Barth cell construction with a particular splitting of cubes into six identical tetrahedra in a Cartesian mesh forming two prisms (Figures 1.4.2 and 1.14). With this combination, the cells are exactly cubes and the scheme is equivalent to the usual vertex-centered-finite volume scheme and the fifth-order accuracy of the V6 scheme holds for advection.

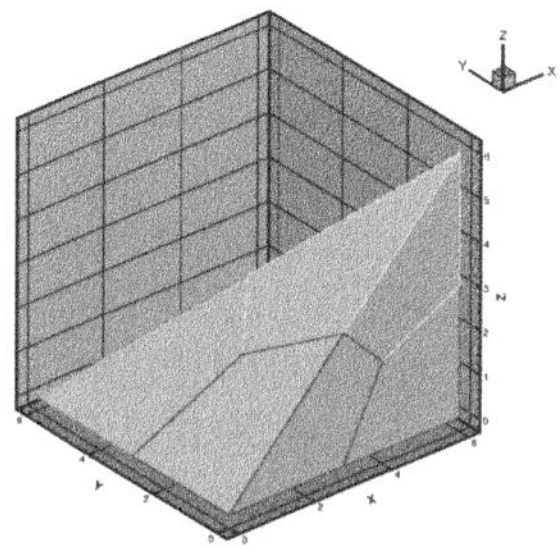

Figure 1.13. *Trace of median cells on a tetrahedron in a particular Cartesian mesh (Gourvitch et al. 2004). For a color version of this figure, see www.iste.co.uk/dervieux/meshadaptation1*

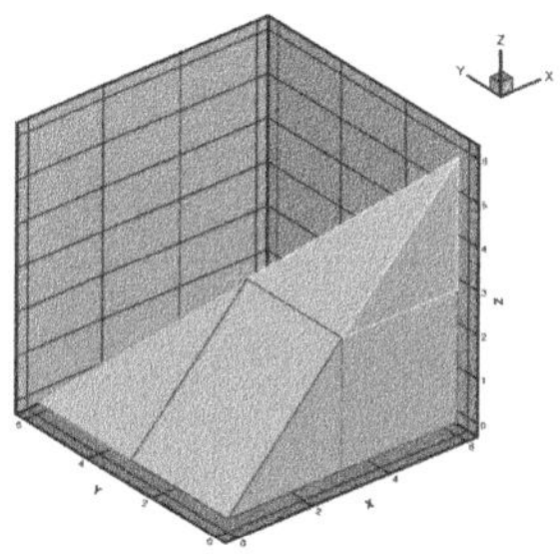

Figure 1.14. *Trace of a circumcenter cell on a tetrahedron in a particular Cartesian mesh (Gourvitch et al. 2004). For a color version of this figure, see www.iste.co.uk/dervieux/meshadaptation1*

1.5. Notes

Both previous models may have solutions fields involving small features challenging for numerical methods. Both flows may involve boundary layers, which are generally much smaller than the mean size of a useable mesh. Compressible flows can have shocks, and multiphase flows show discontinuous interfaces with possibly large change in fluid properties. All these features can be much better addressed when a mesh adaptation is applied, as we shall try to show in this book.

The *upwind-element scheme* described in this chapter was introduced in Stoufflet et al. (1996) for the special treatment of very stiff flows, involving strong bow shocks at large Mach numbers. Positivity statements extend to second-order accuracy the contribution of Farhat et al. (2001).

Further details on the positivity study described in this chapter can found in Cournède et al. (2006).

Interface representation. The numerical representation of an interface in a flow is a difficult problem, which has motivated thousands of publications. In many situation, we need to describe the interface moving in a fixed mesh. Among the methods that can achieve this, we distinguish thick interface techniques (often based on Riemann solvers; see Saurel and Abgrall 1999), interface direct tracking (Unverdi and Tryggvason 1992), volume of fluid (VOF) (Popinet and Zaleski 1999) and level set (LS) (Sethian 1999; Osher and Fedkiw 2002). The issue of the conservation of each phase is well addressed by VOF and not by LS, which has other attracting features (easy extension to high order and better discretization of curvature, among others). The conservative LS of Eça et al. (2007) replaces the distance function by a smooth transition between one phase density and the other one. This transition spreads over a small number of mesh sizes Δx, but this small number needs be sufficiently increased when we try to converge to the exact solution with a reasonable order.

2

Mesh Convergence and Barriers

This chapter identifies two strong advantages of mesh adaptation; *singularity capturing* and *early capturing* of solution heterogeneities allowing, with smaller mesh sizes, an asymptotic convergence of the same order as truncation analysis order. Examples are given with different sorts of mesh adaptors. These two properties are not attained by an arbitrary mesh adaptive strategy. Indeed, most strategies are limited by *convergence barriers*. In particular, whether they are isotropic or anisotropic can change this limitation. We identify several important convergence barriers to which some mesh adaptation families are facing.

2.1. Introduction

While many practical motivations for adapting the mesh are clear, the theoretical advantages of this technique deserve to be well identified. The approximation theory aims at predicting the asymptotic behavior of approximation error when the mesh is refined. For example, the P_1 continuous Galerkin finite element method gives an error four times smaller (in L^2 norm) with a mesh twice finer than an initial mesh, as soon as the initial mesh is sufficiently fine. Let us call it *textbook convergence*. This property can be understood as a complexity property in the sense that it is related to the computational effort to pay in order to get a certain level of accuracy. The second-order accuracy statement cannot be improved to a higher order by applying an adaptive version of this approximation scheme: no mesh adaptation bonus for textbook convergence is when already obtained. The mesh adaptation bonus concerns practical cases where textbook convergence is *not* obtained. In this chapter, we examine an analysis aiming at characterizing the adaptation bonus. It is first presented within an extremely simplified context, the one-dimensional mesh-adaptive interpolation (MAI). In this context, we shall see that some mesh adaptative methods are able to capture 1D discontinuities with an order of accuracy equal to two, while uniform mesh refinement gives only first order. Thus, a global mesh adaptative method, combining an approximation scheme and a well-defined adaptative remeshing, possesses its own convergence order for some family of discontinuities; this convergence order being defined as follows:

DEFINITION 2.1.– *Let u be a function defined on an open subset Ω of the Euclidian space $\mathbb{R}^d$ of dimension d. Let us consider a method approaching in a normed space E the function u, the exact function, with a sequence of discrete functions u_n, described by a number N_n of real numbers, with an error norm $\varepsilon_n = ||u - u_n||_E$. The convergence order of the sequence (u_n) is said to be at least α if we have*

$$\varepsilon_n = ||u - u_n||_E = O(N_n^{-\frac{\alpha}{d}}). \quad [2.1]$$

If for any sequence generated by a discretization method coupling an (adaptative or not) mesh refinement with an approximation or interpolation scheme, the order of the sequence is (at least) α, the discretization method is said of order (at least) α, which we write

$$||u - u_N||_E = O(N^{-\frac{\alpha}{d}}). \square \quad [2.2]$$

Identifying this order of accuracy will rely on several analyses.

– From one side, it is necessary to analyze the accuracy order in the case where the limit of the discretization, that is, the exact solution, is smooth. This analysis can be done with uniform meshes and the usual accuracy order of the approximation scheme is evaluated. The corresponding scheme accuracy order is generally equal to the truncation order of the approximation[1].

– If the continuous limit contains *singularities*, the accuracy order on uniform meshes is lower than in the smooth case. Two typical examples of discontinuities of dimension $d - 1$ are born by a section of curve in 2D and a piece of surface in 3D. In the case of a discontinuity of dimension $d - 1$ and of non-zero measure, the accuracy in $E = L^p(\Omega)$ $(p < \infty)$ is not better than $1/p$.

– In addition to these uniform-mesh analyses, we can take into account in our analysis the application of a *mesh adaptation strategy* and then try to predict its effect on the convergence.

We can interpret a mesh adaptation strategy as a system control strategy. It involves a *sensor* or captor will detect the region of mesh to refine or coarsen, and an *actuator*

1 This may depend on the type of mesh chosen. For example, the accuracy order of an approximation applied to unstructured meshes can be smaller than the accuracy order of the same approximation, restricted to Cartesian meshes, a property known as superconvergence, which is already discussed in Chapter 1.

will build the new mesh. Most of published works in the literature examine the captor, that is, the error estimate. In this chapter, we assume we have a perfect error estimate. Even in that case, many remeshing strategies carry with them a strong limitation of the order of accuracy of the global adaptative approach. More specifically, a typical *barrier*, which will be shown in the sequel, is as follows:

LEMMA 2.1.– If a $\mathcal{P}_1$ interpolation method is applied to a piecewise regular function u, defined on $\mathbb{R}^d$ with $d > 1$, and involving discontinuities of dimension $d-1$ and non-zero measure, then, when the mesh is refined with an adaptative isotropic mode, dividing in the same way Δx, the convergence order α in L^p $(1 \leq p < \infty)$ cannot be better than

$$\alpha \leq \frac{1}{p}\frac{d}{d-1}. \square \qquad [2.3]$$

More precisely, for $p = 2$, L^2 convergence order may be limited to $1/2$ in 2D and $3/4$ in 3D. Such barriers are rather simple to prove and of general application. Indeed, the limitation put in evidence for interpolation extends naturally to PDE approximate solutions. On the contrary, some mesh adaptation strategies allow high-order (> 1) convergence, even on singularities. This property is connected to the ability of *early capturing*, that is capturing the solution details with coarser meshes. In this chapter, early capturing is defined in 1D. Then finite-element multidimensional meshes are considered and the notions of isotropy and anisotropy are introduced. The various approximation properties are analyzed in the simpler case of MAI. Isotropic and anisotropic cases are addressed for the Lagrange interpolation on tetrahedra. After numerical illustrations, the case of unsteady mesh adaptation is considered.

2.2. The early capturing property

2.2.1. *Smoothness, non-smoothness, heterogeneity*

We are interested in *computable non-smooth functions,* that is, which can be discretized with a finite set of real numbers, typically *involving one or several singularities, the support of each of them being a manifold of dimension* $d-1$ *or less, and of measure greater than zero in this manifold.* This is generally the case in numerical simulation. Let us consider the solution u of a PDE. We can say that u is a *highly heterogeneous solution* if it verifies the following:

– (C_1) either u is a smooth function, but its variation contains too small details in comparison with the usual uniform mesh size;

– (C_2) either u is a singular function.

We have presented in Table 2.1 some well-known properties of the approximation error

$$\varepsilon_h = \|u_h - u\| \qquad [2.4]$$

for typical approximation methods, when we apply them to *regular* or *heterogeneous cases*.

We observe in Table 2.1 that the behavior of central finite differences methods, even used in combination with uniform meshes is not satisfactory when they are applied to highly heterogeneous functions. The improvement carried by TVD or ENO *does not* recover the optimal accuracy.

	(1) Regular case	(2) Highly heterogeneous case	(3) Singular case
High order Cent. differ.	$\varepsilon_h = O(\frac{1}{N^{\alpha/d}})$	Oscillations If N not sufficiently large	Oscillations
TVD, ENO (TVD: L^1 error)	$\varepsilon_h = O(\frac{1}{N^{\alpha/d}})$	$\varepsilon_h = O(\frac{1}{N^{1/d}})$ If N not sufficiently large	$\varepsilon_h = O(\frac{1}{N^{1/d}})$

Table 2.1. *Behavior of the approximation error $\varepsilon_h = \|u_h - u\|$ in dimension d when using several approximation methods of potential accuracy $\alpha > 1$ for the resolution of the discrete problem. N is the number of grid nodes. Three cases are considered: (1) regular function, (2) highly heterogeneous function, assuming that some details are too small for the mesh and (3) function with at least one singularity*

2.2.2. Behavior of the uniform-mesh strategy

We first restrict to MAI, which can be defined as follows:

Given a function u, given interpolation operators Π_h defined on a family of meshes $\mathcal{M}_h$, given a norm $\|.\|$, given an integer N, find the mesh $\mathcal{M}_h^{opt}$ of N vertices which minimizes the interpolation error $\|\Pi_h u - u\|$.

MAI issues are met when, before solving a PDE, we build a mesh that needs to be adapted to *known* data and geometry. MAI also concerns data compression and function tabulation. Function $u(x)$ can be well defined but expensive to compute. $\Pi_h u(x)$ is less expensive as far as it is easy to store the description of $\Pi_h u$ and to compute $\Pi_h u$ from it; we shall roughly define *the complexity or cost of* $\Pi_h u$ as the number N of degrees of freedom used to represent $\Pi_h u$. Let u be a function defined on interval $[0,1]$ on which we consider a subdivision h_N:

$$0 = x_0 < x_1 < \cdots < x_{N-2} < x_{N-1} < x_N = 1$$

and the P_1-Galerkin interpolation u_{h_N} of u on this subdivision is as follows:

$$\text{For } x_i \leq x \leq x_{i+1} : u_{h_N}(x) = \alpha u(x_i) + (1-\alpha)u(x_{i+1}) \text{ where} : \alpha = \frac{x - x_{i+1}}{x_i - x_{i+1}}.$$

The P_1-continuous interpolation is known to be second-order accurate for the L^2 norm. This means that a twice finer mesh, such as

$$\begin{aligned}&0 = \tilde{x}_0 < \tilde{x}_1 < \cdots < \tilde{x}_{2N-2} < \tilde{x}_{2N-1} < \tilde{x}_{2N} = 1\\ &\tilde{x}_{2p} = x_p \qquad \tilde{x}_{2p+1} = (x_p + x_{p+1})/2,\end{aligned}$$

gives a four times smaller norm of the error u_h. Unfortunately, this occurs only when the mesh is sufficiently fine to *capture* the smallest details of the function u to be interpolated, and only if u is sufficiently regular. We illustrate these remarks by considering a series of uniform meshes $\mathcal{M}_N$ which involve N nodes, $x_{i+1} - x_i = 1/N - 1$ $i = 0 \cdots N$ and the two following functions:

$$\begin{aligned}u_1 &= -\, sign\left(x - \frac{1}{2}\right)\\ u_2 &= 1 - \frac{exp(2x/\lambda) - 1}{-1 + exp(1/\lambda)}, \quad \text{for } x \leq \frac{1}{2}\\ &= -1 + \frac{exp(-2x + 1/\lambda) - 1}{exp(1/\lambda) - 1} \quad \text{for } x \geq \frac{1}{2} \quad \text{with } \lambda = 0.005.\end{aligned}$$

Note that u_2 is a regularized version of u_1. The behavior of the usual linear interpolation L_2 error for $\|\Pi_h u - u\|_{L^1}$ for uniform meshes as a function of the number N of nodes used is depicted in Figure 2.1 for the above examples (dash lines); we observe the following:

– the interpolation of u_1 shows first-order accuracy for any mesh fineness;

– first-order accuracy is shown initially by the regular case u_2, with a part of the convergence curve very close to the one of u_1: if at most 30 nodes are used, then the interpolation behaves just as a first-order accurate one. After the “investment” of about 100 nodes, the smooth (but steep) step at $x = 1/2$ is properly captured and second-order convergence appears.

We want to stress that these two events are very frequent in numerical analysis, either a singularity spoils the convergence, or there are simply steep gradients, but the meshes chosen stay in the singularity-like first-order convergence.

2.2.3. *An example of 1D adaptation*

Let us give a first example of mesh adaptation which produces second-order convergence. Let us consider more generally that u, defined on an interval $[0,1]$, is regular in $[0,1]$ except in a point I, where it is discontinuous. Given a mesh subdivision of $[0,1]$ with N nodes, let us call V_h the interval containing I and V_h' its complementary in $[0,1]$. The following estimates hold:

$$\begin{aligned} \exists C &> 0, \text{ such that} \\ \varepsilon_h &= \|\Pi_h u - u\|_{L^1(V_h)} \sim \mathrm{C/N} \\ \varepsilon_h' &= \|\Pi_h u - u\|_{L^1(V_h')} \sim \mathrm{C/N^2}. \end{aligned}$$

An adequate adaptive strategy is defined as follows:

LEMMA 2.2.– If we divide each interval of V_h' into two equal intervals, and divide into four equal intervals the interval in V_h, then the L^1 norm $\|\Pi_h u - u\|^*_{L^1([0,1])}$ on the refined mesh and the L^1 norm $\|\Pi_h u - u\|_{L^1([0,1])}$ on the initial mesh are related by

$$\|\Pi_h u - u\|^*_{L^1([0,1])} \sim 1/4\, \|\Pi_h u - u\|_{L^1([0,1])}, \qquad [2.5]$$

for a number of nodes passing from N to $2N+1$ and, asymptotically,

$$\|\Pi_h u - u\|^*_{L^1([0,1])} = O(1/N^2) \qquad [2.6]$$

which is the statement of the second-order convergence. □

DEFINITION 2.2.– *The property of a mesh adaptation algorithm to give a second-order convergence for a singular function is called the singularity capturing property. The property of a mesh adaptation algorithm to early give a second-order convergence for a highly heterogeneous function is called the early capturing (EC) property.*

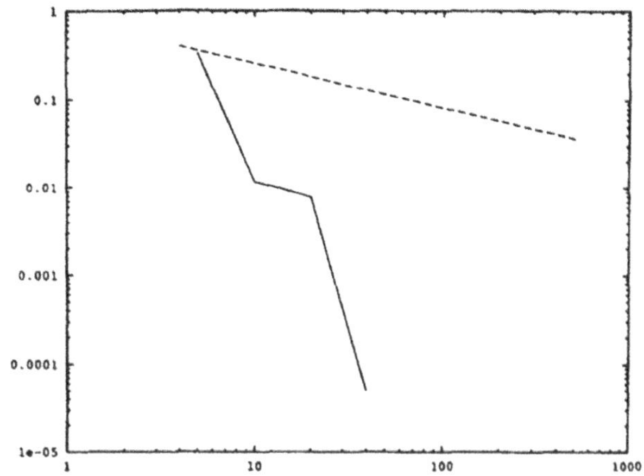

a)

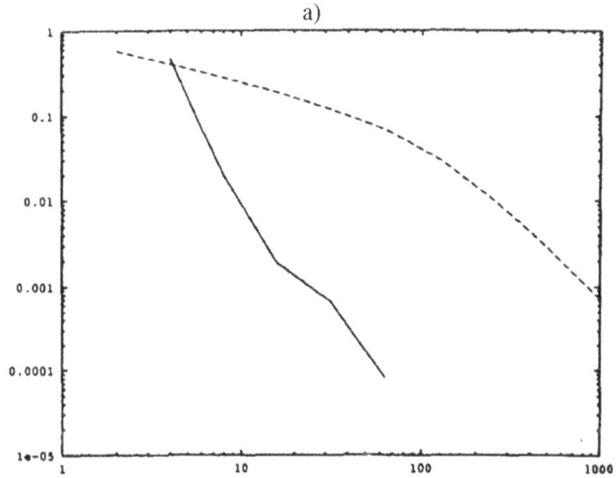

b)

Figure 2.1. *a) Mesh convergence for u_1: the error norm $\|u_1 - u_{h1}\|_{L^1([0,1])}$ along ordinates is a function of the number of nodes (abcissae). In the dashed line, (- - -), uniform mesh convergence order is 1, in the continuous line, (—), the mesh-adapted convergence order is larger than 2. b) Mesh convergence for u_2: the error norm $\|u_2 - u_{h_2}\|_{L^1([0,1])}$ along ordinates is a function of the number of nodes. In the dashed line, (- - -), the uniform mesh convergence presents first a first-order slope (compare to previous figure) and then turns to second-order convergence after the small detail has been well represented – "captured" – by the local mesh. In the continuous line (—), mesh-adapted convergence is, even for coarse meshes, of slope at least 2*

REMARK 2.1.– In Palmerio and Dervieux (1996), the second-order convergence of the well-known spring-analog mesh adaptation model of Gnoffo (1983); Palmerio (1994, 1988, 1994) and Palmerio and Dervieux (1990) is also proved.

2.3. Unstructured meshes in finite element method

2.3.1. *Basics of finite element meshes*

Finite elements meshes are first based on the notion of a single finite element.

DEFINITION 2.3.– *A finite element is defined as the triplet* (K, P_K, Σ_K) *where*

– K *is a geometrical polygonal (2D)/polyhedral(3D) subdomain of the Euclidean space* $I\!R^2$*(resp.* $I\!R^3$*), the element;*

– P_K *is a space of real-valued functions defined on* K*, the space of basis functions;*

– Σ_K *is a set of linear functionals on* P_K*, the degrees of freedom.*

A typical example is the triplet $(K, P_1, \text{values at vertices})$ where K is a triangle. It describes the linear approximation in a triangle defined from values at vertices.

The above general definition needs to be completed by assumptions such as:

– K is not degenerated (of non-zero measure and sufficiently regular);

– any choice of degrees of freedom specifies a unique function of P_K.

A *computational domain* is an open subset Ω of $I\!R^n$ bounded by a (piecewise) regular manifold $\partial\Omega$ of dimension $n-1$, Ω being on one side of $\partial\Omega$.

DEFINITION 2.4.– *We call* τ_h *be an admissible tetrahedrization for finite elements of* Ω *if, for the 3D case,* Ω *is partitioned in tetrahedra, and the intersection of two different tetrahedra is either empty, or a vertex, or an edge, or a face (respectively, triangles with intersection empty, vertex, or edge in* $I\!R^2$*).*

The index h refers to a choice of the size of the elements covering Ω (with smaller sizes, the number of elements will necessary be larger). But for a general mesh, knowing the parameter h, generally chosen as the largest edge, is not sufficient for defining the sizes of all elements. Another point is that, for most cases, only an approximation of Ω, the *numerical computational domain* Ω_h is covered by τ_h, and to a family of meshes $(\tau_h)_{h\in\mathcal{H}}$ generally corresponds a family $(\Omega_h)_{h\in\mathcal{H}}$, which should better approximate Ω when h is decreased.

2.3.2. *Anisotropy*

We consider the interpolation of more or less smooth functions defined in a regular bounded domain Ω of $\mathbb{R}^d$ ($d = 1, 2$, or 3). The interpolation methods under study are designed from a discretization of the computational domain Ω, the mesh, τ, which relies on a partition of Ω into N polygonal elements, which we denote K, and that are assumed to be convex, open sets and such that $\cup_\tau \bar{K} = \bar{\Omega}$. We examine the convergence properties of an interpolation method over a sequence of meshes $(\tau_p)_{p\in\mathbb{N}}$. We assume that in this sequence of meshes, the size of elements tends to zero. Accordingly the number of elements tends to infinity.

DEFINITION 2.5 (Diameter and roundness).– *For any polygon K in $\mathbb{R}^d$, its diameter h_K is the diameter of the smallest sphere containing it and its roundness ρ_K is the diameter of the largest sphere inside it.*

The ratio ρ_K/h_K measures the shape of K : when it is close to 1, K is close to a sphere or nearly isotropic, when it is close to 0, K is much stretched or strongly anisotropic. In the case of triangles, the maximal ratio is $1/2$ and is reached for an equilateral triangle. In the case of tetrahedra, the maximal ration is $1/3$ and is reached for an equilateral (i.e. regular) tetrahedron.

DEFINITION 2.6 (Quasi-uniform meshes sequences).– *A sequence of meshes $(\tau_p)_{p\in\mathbb{N}}$ is quasi-uniform if*

$$\exists\, c > 0, \quad \forall p \in \mathbb{N}, \quad c \max_{\tau_p} h_K \le \min_{\tau_p} \rho_K .$$

Then $h_p = \max_{\tau_p} h_K$ can be considered as a measure of the mesh size of τ_p.

In a sequence of uniform meshes, c reaches a maximal value for regular elements (e.g. $1/3$ for tetrahedra) and no element has a roundness smaller than c. The elements are essentially of same size and roundness. The global mesh fineness or size is defined roughly by a single number. In other words, the number of elements N_p is related to the size h_p

$$\exists\, c^\star > 0, \quad \forall p \in \mathbb{N}, \quad c\,h_p^d \le c^\star \frac{\text{meas }(\Omega)}{N_p} \le h_p^d, \qquad [2.7]$$

since $\rho_K^d \le c^\star \text{meas }(K) \le h_K^d$ for any K (c is the constant arising in definition 2.6). This can be also written as:

$$h_p = O\left(N_p^{-\frac{1}{d}}\right).$$

DEFINITION 2.7 (Mesh isotropy).– *A sequence of meshes* $(\tau_p)_{p\in\mathbb{N}}$ *is isotropic if it satisfies*

$$\exists\, c > 0, \quad \forall p \in \mathbb{N}, \quad c \leq \min_{\tau_p} \frac{\rho_K}{h_K}.$$

This property prevents an infinite stretching, that is, prevents the ratio ρ_K/h_K from tending to zero 0. However, in contrast to uniform meshes, in an arbitrary isotropic mesh the elements can be of very different size. In practice, *a mesh is isotropic* when each of its elements satisfies the above inequality with a ratio c rather close to the maximal ratio (regular element).

REMARK 2.2.– Two notions are mainly related to mesh quality and need to be taken into account in mesh regeneration. *Mesh size variation* refers to the variation of local characteristics (element size, directions of stretching, strengths of stretching) between two neighboring locations of the computational domain. *Mesh gradation* masters this variation between neighboring elements. *Connectivity variation* for unstructured meshes refers to the variation of the number of neighbors for element or vertices.□

From now on, we restrict to meshes involving simplexes, that is, in 2D, triangles, and in 3D, tetrahedra.

2.4. Accuracy of an interpolation

In the standard theory, the approximation error is measured as a function of a representative (largest) mesh size h. This notion is well adapted to uniform sequences of meshes. It extends also to sequences that are uniformly refined. In mesh adaptation methods, the meshes are not uniform, and their sequences are not uniformly refined. In the classical approximation theory, the mesh fineness is a measure of the cost to be paid in order to compute the considered mesh. The idea of simply considering the total number of degrees of freedom for adapted meshes is a natural extension. We illustrate this for interpolation.

We focus on the interpolation of functions of $H^k(\Omega) \cap L^\infty(\Omega)$ with $k \in \mathbb{N}$. An interpolation method is a mapping that defines from any mesh τ of an open set Ω in $\mathbb{R}^d$, an operator $\Pi_\mathcal{T}$ such that

$$\Pi_\tau : u \in H^k(\Omega) \cap L^\infty(\Omega) \longrightarrow \Pi_\tau u \in L^\infty(\Omega).$$

Let us consider the interpolated functions over a sequence of meshes $\tau_1, ..., \tau_p, ...$: It is useful to introduce the following property:

$$\begin{aligned} &\exists\, c > 0, \quad \forall p \in \mathbb{N},\ \forall u \in H^k(\Omega) \cap L^\infty(\Omega) \\ &\|\Pi_{\tau_p} u\|_{L^\infty(\Omega)} \leq c \min(\|u\|_{H^k(\Omega)}, \|u\|_{L^\infty(\Omega)}), \text{ and} \\ &\|\Pi_{\tau_p} u - u\|_{L^2(\Omega)} \to 0, \text{ when } p \to \infty. \end{aligned} \qquad [2.8]$$

Two classical examples satisfy this property:

– the Lagrange interpolation on triangular or tetrahedral meshes;

– the L^2-orthogonal and piecewise constant projection (for any kind of meshes).

The first example is a usual one in finite-element theory, while the second one is adapted to finite-volume methods.

DEFINITION 2.8.– *An interpolation method Π_{τ_p} has an order of accuracy of (at least) α for a sequence of meshes $(\tau_p)_{p\in\mathbb{N}}$ if*

$$\exists\, c > 0, \quad \forall u \in H^k(\Omega) \cap L^\infty(\Omega),\ \forall p \in \mathbb{N}, \quad \|u - \Pi_{\tau_p} u\|_{L^2(\Omega)} \leq c N_p^{-\alpha/d},$$

where N_p holds for the number of elements of τ_p.

2.5. Isotropic adaptative interpolation

While, according to finite element convergence theory (e.g. (Ciarlet 1978)), isotropy is *a priori* better for accuracy in an element, this section shows that isotropy is not the right strategy for mesh adaptation. We consider sequences of adapted meshes for functions having discontinuities. We successively address the 1D case, isotropic adapted meshes (see definition 2.7) for the 2D case and the 3D case. We show convergence barriers by exhibiting counterexamples.

2.5.1. *The 2D case*

In order to evaluate the order of convergence according to [2.1] and [2.2], we introduce the following computational lemma:

LEMMA 2.3.– We assume that when for passing from an interpolation error ε to an interpolation error ε/q, the number of degrees of freedom (d.o.f.) is multiplied by β. Then convergence order is $\alpha = log(q^d)/log(\beta)$.

Indeed, starting from the bound $\varepsilon \leq KN^{-\alpha/d}$, we can write that $\frac{1}{q} = (N/(\beta N))^{-\alpha/d}$.□

We introduce now the following two-dimensional Heaviside function:

$$For\ (x, y) \in]0,1[^2,\ H(x, y) = \begin{cases} 0 & \text{if } x < 0.5 \\ 1 & \text{if } x > 0.5 \end{cases}$$

On each element of a mesh containing the discontinuity, the integral of square of interpolation error $\varepsilon^2(x)$ is just $O(1)$ times the element area, passing from a uniform mesh to a second uniform mesh with a mesh size twice smaller divides by two the total integral of $\varepsilon^2(x)$ and by $\sqrt{2}$ the L^2 norm. Convergence order is $\frac{1}{2}$ ($\frac{1}{p}$ pour L^p), close to discontinuity. It is a global convergence order $\frac{1}{2}$ ($\frac{1}{p}$ pour L^p) for uniform meshes with the previous lemma with $\beta = 4, q = \sqrt{2}, d = 2$. In the case of an isotropic mesh adaptation, we have the following:

LEMMA 2.4 ($d = 2$).– Convergence in L^2 of the P_1 interpolation of the Heaviside function H is at best of order $\frac{1}{2}$ for *uniform meshes*, and 1 for isotropic meshes (for $L^p, 1 \leq p < \infty$, respectively, $log(p)/(2log(2))$ and $log(p)/log(2)$).

The idea is that the error is concentrated along the discontinuity (Figure 2.2). It is sufficient to evaluate the mesh increment on discontinuity. Since the error close to discontinuity decreases as $\Delta x^{\frac{1}{2}}$, to divide the error by a factor $\sqrt{2}$, we need to multiply the number of d.o.f. on discontinuity by 2. Then $\beta = 2, q = \sqrt{2}, d = 2$ and therefore $\alpha = 1$.□

2.5.2. *A first 3D case*

The above calculation can be transposed to a singularity situated on a straight line in 3D. A typical example is a creek in solid mechanics (see, for example, a mesh adapted computation in Tie and Aubry 1999).

LEMMA 2.5 ($d = 3$).– For a function u defined on $\mathbb{R}^3$, having a singularity on a straight line, the best order of accuracy in L^p for isotropic adaptive strategies is $(3/2)log(p)/log(2)$.□

The error is concentrated along the discontinuity (Figure 2.3). Since the error close to discontinuity decreases as $\Delta x^{\frac{1}{2}}$, in order to divide the error by a factor $\sqrt{2}$, we need to multiply the number of d.o.f. on discontinuity by 2. Then $\beta = 2, q = \sqrt{2}, d = 3$ and therefore $\alpha = 3/2$.□

Figure 2.2. *Isotropic adaptation for a (vertical) discontinuity. For dividing the L^2 interpolation error by a factor $\sqrt{2}$, twice as many degrees of freedom are necessary. The figure presents the building of a new mesh by isotropic division when square elements are used*

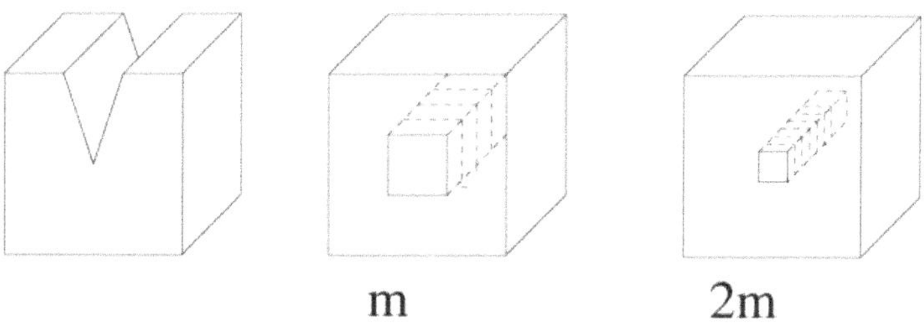

Figure 2.3. *Generation of a new adapted mesh in the case of a creek. Left: The edge of the dihedron bears the creek, a singularity which needs local refinement. Center and right: A better capture is obtained by refining along this edge at the cost of solely doubling the degrees of freedom*

REMARK 2.3.– Conversely, uniform refinement strategy would have required N^3 nodes for the same level of error, that is to say, only a first-order accuracy would be obtained.

2.5.3. *A limiting barrier for the isotropic 3D case*

In three-dimensional CFD, the most important singularity is the shock wave, which is a discontinuity on a surface.

LEMMA 2.6 ($d = 3$).– Let H be the 3D Heaviside function on $]0, 1[^3$ (0 for $x < \frac{1}{2}$ and 1 on $x > \frac{1}{2}$). Then the maximal order of convergence of a mesh adaptative isotropic P_1-interpolation method in L^p is $(3/4)log(p)/log(2)$.

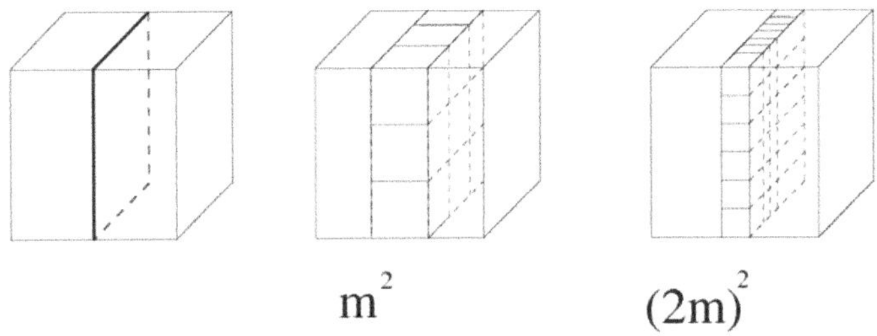

Figure 2.4. *Left: The singularity is born by a vertical plan. Center and right: A new mesh is generated by local isotropic division in 3D*

PROOF.– For dividing the error by $q = \sqrt{2}$, we need to multiply the d.o.f. by $\beta = 4$ then $\alpha = 3/4$ (lemma 2.3)[2]. □

2.6. Anisotropic adaptative interpolation

For P^1 *anisotropic* interpolation, the above counter-examples cannot be stated. At the contrary, we shall see in many experiments that the higher order accuracy of the interpolation is preserved by a class of anisotropic mesh adaptations. We propose in this section the analysis of a typical example. Indeed, for the interpolation of a multidimensional Heaviside function, some arguments can be exhibited, which tend to show that P_{k-1} anisotropic interpolation can achieve kth order discontinuity capturing in many cases.

2.6.1. *Anisotropic adaptation of a Heaviside function*

We consider the same multidimensional Heaviside function $H(x, y, z)$ as in defined in lemma 2.6. The family of meshes which we consider is defined as follows:

2 For the case of a discontinuous gradient, for example $u = (x - \frac{1}{2})H(x)$, the analysis gives $(3/2)log(p)/log(2)$

– the vertices are generated by three families of parallel plans, orthogonally to Ox,Oy,Oz;

– the plans orthogonal to Oy and Oz are equidistant, that is, the mesh size in direction Oy is equal to the mesh size in direction Oy;

– for simplicity, we assume that the discontinuity plan $x = \frac{1}{2}$ is strictly between two mesh plans $x = x^-$ and $x = x^+$, orthogonal to Ox.

The union of the triangles between these two plans is called singularity-covering regions (SCR). We condider a mesh of the above family, with Oy and Oz common mesh size equal to Δy. The number of degrees of freedom is $N = N_x \times N_y \times N_z$. A finer mesh, still in this family, is defined as follows:

– Oy and Oz common mesh size are equal to $\Delta y/2$;

– outside of the SCR, Δx is divided by two; in other words, any tetrahedron of the SCR is regularly divided into eight tetrahedra;

– inside the SCR, the interval $]x^-, x^+[$ is divided into kp intervals.

As concerns the resulting interpolation error, it remains zero in the complement of the SCR (it would be divided by 2^k if our Heaviside separates two smooth functions), while it will be divided by 2^k in the SCR, since in 1D the convergence order in L^p of any P_k interpolation on a discontinuity is $\frac{1}{p}$. As concerns the increment in d.o.f, N_y becomes $2N_y$,N_z becomes $2N_z$. If SCR had been divided as the rest of mesh, N_x would have become $2N_x$. Instead of adding a single plan (orthogonal to Ox), the adaptation has added $7k$ new plans, in such a way that N_x becomes $2N_y + 7k - 1$. Therefore, the number of d.o.f. is increased from $N = N_x \times N_y \times N_z$ to $8\kappa N$ with

$$\kappa = \frac{1}{8}\frac{(2N_x + 7k - 1) \times 2N_y \times 2N_z}{N_x \times N_y \times N_z}.$$

Since the new mesh also belongs to the above mesh family, this process can be reiterated m times, resulting in the convergence of N_m,$N_{x,m}$, $N_{y,m}$,$N_{z,m}$ to infinity. For these integers sufficiently large, we observe that

– the interpolation error is divided by 2^k at each iteration:

$$||\Pi_{k-1}^{\mathcal{M}_m} H - H||_{L^p} \leq 2^k ||\Pi_{k-1}^{\mathcal{M}_{m-1}} H - H||_{L^p};$$

– the number of d.o.f. is increased by a factor $\kappa_m 2^3$, where $\kappa_m \to 1$ when m tends to infinity. Using lemma 2.3 with $q = 2^k$, $\beta = 8$, expresses that the interpolation error of H on this sequence of meshes converges at an order k as follows in lemma 2.7:

LEMMA 2.7.– In the case of a multidimensional Heaviside function H, there exists sequences of isotropic adapted meshes $\mathcal{M}_m$ involving N_m vertices, which, combined with a P_{k-1} interpolation, yield kth order convergence in $L^p, 0 \leq p < \infty$:

$$||\Pi_{k-1}^{\mathcal{M}_m} H - H||_{L^p} \leq K N_m^{-\frac{k}{d}}. \square$$

As noted in the above proof, this results extends to the case of a planar discontinuity separating two smooth functions. The 2D case holds with discarding N_z.

2.6.2. *Heaviside function with curved discontinuity*

Let us now assume that the singularity is a smooth curve with bounded curvature. We deform the family of adapted mesh accordingly, passing from orthogonal (x, y, z) to curvilinear (ξ, η, ζ), while keeping straight elements (straight edges). The discontinuity surface can be locally considered as a paraboloid. In order to allow the surface to stay inside the thickness of an element, the mesh size normal needs be proportional to the square of mesh size tangent. This allows to contain the curve in a row of elements. Let us investigate how to get an error 2 times smaller ($q = 2$):

– as concerns the region out of SCR, it is sufficient to divide the mesh size by two;

– in the SCR, we need to divide by a factor 2^p normally to discontinuity and *due to the curvature* by a factor $(2^p)^{\frac{1}{2}} = 2^{\frac{p}{2}}$ in the two directions tangent to the discontinuity.

Conformal element matching is applied at the boundary of SCR. The number of nodes may pass

– from N_{out} to $8N_{out}$ out of the SCR;

– from N_{SCR} to $N_{SCR} * 2^{kp} * 2^{\frac{kp}{2}} * 2^{\frac{kp}{2}}$, with a new SCR defined as the part of the divided SCR, which is crossed by the discontinuity.

But in the new SCR, we may have only one layer normally to the discontinuity and the number of nodes of the new SCR is $N_{SCR} * 2^{\frac{p}{2}} * 2^{\frac{p}{2}}$, then $\beta = 2^p$. We concentrate as in previous isotropic analyses on the asymptotically dominant increment on the SCR, and get an order of $\alpha = \frac{3}{p}$ for the singular convergence independently of the truncation error k of the interpolation, and globally, an order of $\alpha = \min(k, \frac{3}{p})$. In 2D, the new SCR involves $N_{SCR} * 2^{\frac{p}{2}}$ vertices, then $\beta = 2^{\frac{p}{2}}$.

LEMMA 2.8.– In the case of a multidimensional curved Heaviside function H, the convergence order is at best $\alpha = \min(k, \frac{3}{p})$ in 3D and $\alpha = \min(k, \frac{4}{p})$ in 2D. $\square$

2.7. Numerical illustration: anisotropic versus isotropic interpolation

The above theoretical analyses may have an important practical impact. In particular, it is very easy to experience the above barriers by direct application of a mesh adaptative strategy. For illustrating this point, we use a modern mesh adaptation algorithm defined in the following chapters. It consists of a loop of successive mesh adaptations with a prescribed number of nodes. New meshes are triangulations generated according to metrics derived from an approximation of the Hessian (at previous mesh nodes). The problem addressed is defined as follows. We want to interpolate as accurately as possible in L^2 with the continuous piecewise P_1 operator and a unique adapted mesh of a square, two given discontinuous functions. The Heavyside functions H_1 and H_2 are discontinuous, respectively, on $x = 0$ for the first one and on $y = 0$ for the second one. For a prescribed number n of vertices, we obtain a L^2 error $\mathcal{E}_n$. We compute this error for various mesh fineness n, from 60 to 20,000 with three families of meshes:

– The first family contains uniform meshes.

– The second family of calculations has been done with isotropic mesh adaptation. The BAMG adaptator (Castro-Díaz et al. 1996) is used with a scalar metrics defined on any vertex of the current mesh and chosen as the maximum of the absolute values of the eigenvalues of the local approximate Hessian of the two functions to interpolate. Each mesh is obtained for a prescribed number of vertices when the adaptation loop produces successive meshes that are very close to each other. A typical mesh is presented in Figure 2.5.

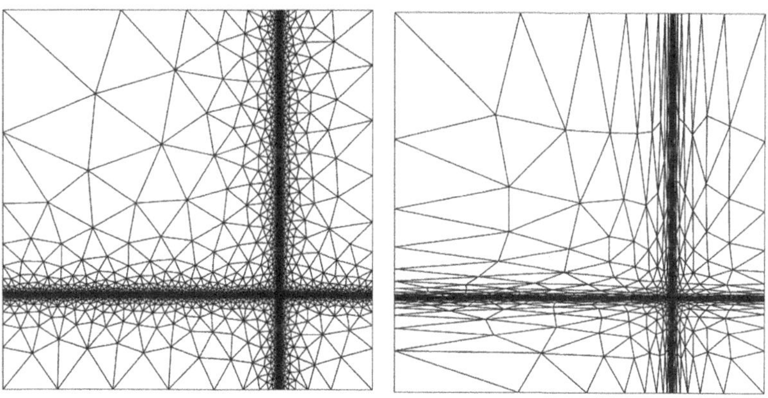

Figure 2.5. *Left: isotropic adapted mesh, right: anisotropic mesh for the two-Heavyside interpolation test case*

– The third family of meshes is generated by anisotropic adaptation. The metric is a matrix, defined on any vertex, obtained from the discrete Hessian by replacing eigenvalues by their absolute values. The synthesis between the two functions to interpolate follows the method proposed in BAMG, computing an ellipse inside the two stretching ellipses (see Castro-Díaz et al. 1996). Again the adaptation iteration loop is reiterated until the successive meshes do not show much variation. A typical example of adapted-converged mesh is presented in Figure 2.5, right, and the convergence order is presented in Table 2.2.

Mesh strategy:	Uniform	Isotropicly adapted	Anisotropicly adapted
Convergence:	$\approx$1/2	$\approx$1	$\approx$2

Table 2.2. *L^2 convergence order observed for the two-Heaviside interpolation test case*

2.8. CFD applications of anisotropic capture

The examples that are presented now concern the adaptation of a PDE solution. The adaptation algorithms are described in details in the following chapters. We concentrate here on their behavior with respect to the convergence barriers defined in the present chapter.

2.8.1. *Pressure with discontinuous gradient*

Let us take as example a typical problem in multiphase flow related with the discontinuity of pressure gradient along an interface. Our model problem is a Poisson problem with a discontinuous coefficient, the discontinuity follows a circle included in the computational domain (Figure 2.6). We apply three mesh convergence strategies: (1) a uniform refinement and two adaptive strategies defined in next chapters, namely (2) a feature-based adaptation (Chapter 5) and (3) a norm-oriented mesh adaptation (Chapter 6). With both mesh adaptive approaches, second-order convergence is obtained. In contrast, uniform mesh refinement produces a disastrous first-order convergence with a 10% error when using a 300×300 mesh.

2.8.2. *Scramjet flow*

In the case of a third-order accurate interpolation or, for a CFD calculation, in the case of an approximation of order 3, like a quadratic central ENO scheme, a discontinuity has influence on a thicker set of approximation molecules. Let us consider the 1D case: Our error estimate involves a third-order finite difference, weighted by $1/\Delta x^3$ with a support width of a few Δx. The discontinuity will infer a concentration of nodes proportional to $1/\Delta x^3$, stronger than in the second-order case. Limiters can be used. Accuracy returns to first order through the discontinuity, but the shock structure is thinner, therefore easier to capture with a small number of nodes.

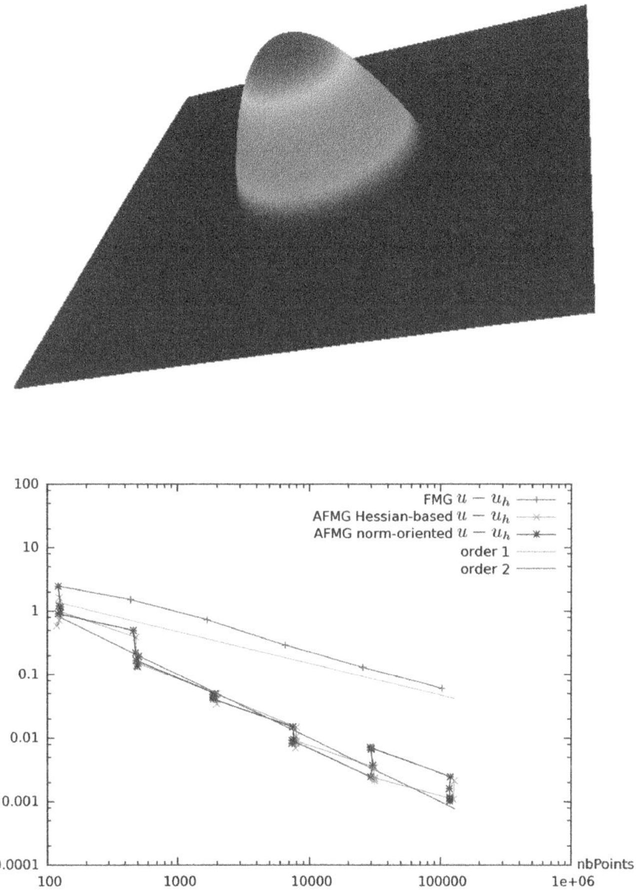

Figure 2.6. *Mesh-adaptive calculation of a pressure with discontinuous gradient due to an interface. The second-order mesh convergence (in terms of number of vertices) of a feature-based adaptation and a norm-oriented adaptation are compared with the convergence of uniform refinement, which is only first-order. For a color version of this figure, see www.iste.co.uk/dervieux/meshadaptation1*

Moving on to multidimensional case, let us examine the configuration where the exact solution involves a discontinuity aligned with a rather regular surface.

Geometrically, the discontinuity can be put between two lines(2D)/surfaces(3D) of nodes. This results in a too large number of nodes if the mesh is isotropic, but a smaller number of nodes (allowing second-order convergence) if the mesh is with straight edges and anisotropic, and finally a reasonable number of nodes (allowing higher order convergence) if the mesh is anisotropic and the edges are curved.

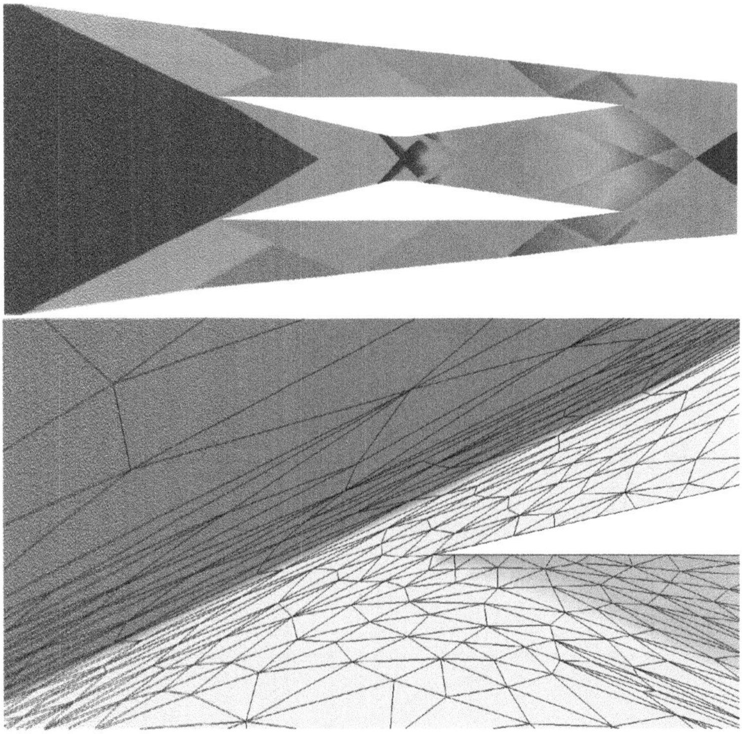

Figure 2.7. *Mesh-adaptive third-order accurate calculation of a scramjet flow, Mach contours, global view and zoom around the bottom leading edge (improved palette). The approximation is of quadratic ENO type combined with flux limiters. This allows a third-order accuracy on regular regions and a thin numerical shock structure (small transition along the diagonal discontinuity). The error estimate does not take into account the limiters and is thus made of third-order derivatives weighted by $1/\Delta x^3$. This induces a high concentration near the discontinuity producing global third-order accuracy. For a color version of this figure, see www.iste.co.uk/dervieux/meshadaptation1*

Due to $(k+1)$th order accuracy far from the singularity, a sequence of mesh which divides by two the mesh size in the regular part of the mesh should divide by at least 2^{k+1} the elements crossed by the discontinuity where convergence accuracy is less or equal unity. Now, according to the type of high-order approximation, the numerical capturing will be spread over one or several high-order elements. Then a refinement region covering a time interval may cover several times 2^{k+1} elements. It may become fairly computational cost consuming. Again, an obvious way to reduce this cost is to use an approximation with one cell per element, like the above referred ENO scheme (Carabias et al. 2018), and to combine it with a shock-capturing mechanisms, which reduces locally the order to one and essentially limits the numerical shock to be spread between only two continuous lines (2D)/surfaces (3D) of cells.

This strategy is applied for obtaining a steady scram jet flow (Figure 2.7) with adaptive capturing of the shocks as thin (in terms of width and of cells) as possible. The scheme and adaptation are those described in Chapter 8 of Volume 2 except they have been simplified here for addressing a steady case. This test case does not involve curved discontinuities for which our adapted mesh with straight edges would face an accuracy barrier (accuracy limited to two). From this standpoint, the path for a general third-order mesh adaptation indeed involves the mastering of curved elements (see, for example, Abgrall et al. 2012).

REMARK 2.4.– In the general case of the Finite Element approximation of PDEs, error estimates can in many cases be expressed in terms of interpolation errors. Then the above barriers should also apply. It is possible to extend the present remarks to a singularity that lies into the *non-boundedness of the domain*. A typical case is the flow around an airfoil. In such a case, it is evident that the domain cannot be meshed uniformly without an infinity of elements. If a truncation before infinity is applied in order to have a bounded mesh, then uniform refinement without enlarging the mesh is also an inefficient strategy since the error of domain truncation can become dominant. In contrast, a reasonable mesh refinement strategy is adaptative with respect to the behavior of the solution at farfield when a very large computational domain is chosen once for all.

2.9. Unsteady case

The *unsteady case*, involving space-time interpolation and space-time simulation, also sets problems of efficiency for discontinuity capturing. As for steady case, the case of the approximation of a PDE cannot be solved with a better accuracy than the case of an interpolation and our discussion will first concentrate on the interpolation case. A preliminary remark is that unsteadiness increases of one unity the dimension of the approximation or interpolation domain. Then barriers are necessarily even more severe. In fact, the only strategy that might avoid a barrier is a "fully space-time adaptative anisotropic" strategy, which is hardly compatible with a *discretization using time levels*. If we consider strategies with time levels, it is useful to distinguish whether the time-step is adapted or not. Further, the best approach with time levels will use a different mesh, adapted, anisotropic at every time level. In that case, due to the time leveling, there will remain a barrier at some order. The popular method of AMR has a different mesh, adapted at each time step, but the spatial adaptation is of isotropic type. It is then asymptotically less good than a *fully space-time adaptative isotropic strategy* and limited by the barrier applying to isotropic approaches. An important particular case is the anisotropic-space/time-leveled cases in which the time discretization is made of time steps $t_1, t_2, ..., t_{nmax}$, while each time level is spatially discretized with an anisotropic unstructured mesh.

2.9.1. *Barriers for second-order time-leveled case*

2.9.1.1. *Unsteady flow with a moving shock*

Let us assume that for each time level, we apply a metric-based anisotropic spatial mesh adaptation, for which the space-wise second-order capture of the singularity is well addressed by the spatial mesh adaptation. Mesh size normal to the discontinuity will be reduced by a factor 4 only in the vicinity of the discontinuity. In other directions near discontinuity and in any direction in other regions of the computational domain, the mesh size is divided by 2. In 3D, with this strategy, passing from N spatial degrees of freedom to $8N$ leads to a four times reduction of the spatial error. Unfortunately, the time step also needs to be reduced by a factor 4, leading to a total number of degrees of freedom multiplied by 32. Thus,

$$||u - u_{32N}||_{L^1(\Omega\times[0,T])} \leq \frac{1}{4}||u - u_N||_{L^1(\Omega\times[0,T])} . \qquad [2.9]$$

By comparing $\frac{1}{4}$ with $32^{-\alpha/n}$, this time with $n = 4$ (space-time dimension), we get the following barrier convergence lemma:

LEMMA 2.9 (Space-time barrier for time advancing discretizations).– For a time-advancing second-order flow solver coupled with an unsteady mesh adaptation at each time level applied to a traveling discontinuity, the space-time convergence rate is at most $\alpha = 8/5$. □

REMARK 2.5.– The limitation at rate $8/5$ applies to the cases where both smooth regions and discontinuities are present. In the case of a flow involving only discontinuities and constant regions, the spatial mesh size can be limited to $4N_{tot}$ and the second order can be recovered. In the case without discontinuity, the second order is automatically recovered because the time step is only reduced by a factor 2. □

In this monograph, we shall focus on time-leveled mesh-adaptative algorithms in which the adapted mesh is frozen during a time sub-interval, made of several time steps (transient fixed point, Chapters 2, 7, 8 of Volume 2). This frozen spatial mesh needs to account for the *motion* of any singularity in the solution. If the singularity, for example a shock wave, moves during the simulation, then the *refined region* of the spatial mesh should cover the *trajectory* of the discontinuity, which is the *union* of all singularity neighboring regions for any time level of the time sub-interval. In contrast to the steady case, this region is a *thick* region. At convergence, the measure of the thick trajectory swept by the moving discontinuity is essentially proportional to its normal velocity times the time sub-interval width ΔT. For a single time sub-interval of fixed width, this measure cannot tend to zero, hence a large number of nodes is wasted. When we apply a process of convergence to the continuous, we divide by 4 the mesh size normal to discontinuity and by 2 in the other directions. With an unchanged time sub-interval width, the number of nodes in the trajectory of the discontinuity would be 16 times larger and 8 times larger for the rest of the mesh. It is thus mandatory to decrease time sub-interval width ΔT by a factor 2 (i.e. to increase their number n_{adap} accordingly in order to recover higher order convergence). We can analyze the spatial convergence order of this strategy by considering only the mean number of nodes, quotient of total space-time number of nodes by the number of time steps: $N_{mean} = N_{st}/n_{\mathcal{T}}$. We get a necessary condition for the spatial convergence order:

LEMMA 2.10 (Spatial convergence for the transient fixed-point mesh adaptation based on a second-order scheme).– For an unsteady flow field with a moving shock, a necessary condition for L^1 -second-order convergence in space is that the time sub-interval width is divided by 2 when smooth region mesh size is divided by 2 in each of the three space directions and when singular region mesh size is divided by 2 in directions parallel to shock and by 4 in direction normal to shock. □

To summarize, in the case of a uniform division into time sub-intervals of size $\Delta T = T/n_{adap}$, we expect to get a *spatial* second-order spatial convergence in $\mathbf{L}^1([0,T];\mathbf{L}^1(\Omega))$ by passing from (N_{mean}, n_{adap}) to $(8N_{mean}, 2n_{adap})$. Indeed, in such a case, the density of nodes is locally multiplied by 16 in the thick trajectory of the discontinuity but the thickness of this region is divided by 2 leading to a total number of nodes multiplied by 4, the main mesh effort comes from the smooth region.

A more realistic evaluation of space-time convergence order needs to account for the number or equivalently the size of the time step. According to lemma 2.9, we cannot hope a convergence at order better than $8/5$. We divide the time step by 4, since the traveling discontinuity is also a timewise discontinuity to be captured accurately with first-order time accuracy. This increases by a factor 4 the space-time number of nodes, which needs also to be increased by a factor 8 at each time level. We also need to divide the time sub-interval length ΔT by 2 for limiting the number of nodes in the vicinity of the discontinuity trajectory.

LEMMA 2.11 (Space-time convergence for the transient fixed-point mesh adaptation based on a second-order scheme).– We consider a second-order flow solver coupled with an unsteady mesh adaptation applied to a 3D traveling discontinuity. We assume a uniform division of the simulation time frame into time sub-intervals of size $\Delta T = T/n_{adap}$. Then, we expect a space-time convergence in $\mathbf{L}^1([0,T];\mathbf{L}^1(\Omega))$ at order $8/5$ by dividing the time step by a factor 4: $\tau \rightarrow \tau/4$ and passing from (N_{st}, n_{adap}) to $(32N_{st}, 2n_{adap})$. □

REMARK 2.6.– Between sub-intervals, transfers of the solution fields from the previous mesh to the new one are necessary. The choice of the transfer operator has certainly some impact on the global accuracy (see, for example, Alauzet and Mehrenberger 2010b) together with how frequently it is applied. The study by Alauzet and Mehrenberger (2010b) suggests to consider conservative interpolation for compressible flows. □

2.9.2. *Barriers for third-order time-leveled case*

The space-wise third-order capture of the singularity can be addressed by spatial mesh adaptation: for each time level, we apply a metric-based anisotropic spatial mesh adaptation. For third-order discontinuity capturing, mesh size normal to the discontinuity in the close vicinity of the discontinuity needs to be reduced by a factor 8. In other directions near discontinuity and in any direction in other regions of the computational domain, the mesh size is divided by only 2. Then the increment in number of nodes near the discontinuity is asymptotically negligible and in total, passing from N spatial degrees of freedom to $4N$ leads to a eight times reduction of the spatial error. But in order to capture *curved discontinuities* with third-order, curved elements seem mandatory. Even in that case, we shall show that third-order time-leveled mesh adaptation is facing rather severe barriers. In 2D, for decreasing accordingly time discretization error, the time step also needs to be reduced by a factor 8, leading to a total number of degrees of freedom multiplied by 32. Thus,

$$||u - u_{32N}||_{L^1(\Omega\times[0,T])} \leq \frac{1}{8}||u - u_N||_{L^1(\Omega\times[0,T])}\,. \qquad [2.10]$$

By comparing $\frac{1}{8}$ with $32^{-\alpha/d}$, this time with $d = 3$ (space-time dimension), we get a space time convergence order $\alpha = 9/5$. In 3D, with this strategy, passing from N spatial degrees of freedom to $8N$ leads to a eight times reduction of the spatial error. Unfortunately, the time step also needs to be reduced by a factor 8, leading to a total number of degrees of freedom multiplied by 64. Thus,

$$||u - u_{64N}||_{L^1(\Omega\times[0,T])} \leq \frac{1}{8}||u - u_N||_{L^1(\Omega\times[0,T])}\,. \qquad [2.11]$$

By comparing $\frac{1}{8}$ with $64^{-\alpha/n}$, this time with $n = 4$ (space-time dimension), we get $\alpha = 2$.

LEMMA 2.12 (Space-time convergence for transient fixed-point mesh adaptation based on a third-order scheme).– For a time-stepping third-order flow solver coupled with an unsteady mesh adaptation at each time level applied to a traveling discontinuity, the space-time convergence rate is at most $\alpha = 9/5$ for 2D unsteady calculations and $\alpha = 2$ for 3D unsteady calculations. □

REMARK 2.7.– We give in Chapter 8 of Volume 2 an example of such third-order scheme.

2.10. Conclusion

This chapter concentrates on the approximation of two particular types of functions.

First, we define a particular class of *non-smooth functions*. They have very small subregions of singularity, born by rather regular manifolds. Of course, the difficulty in approximation depends strongly on the nature of the singularity. The results of this chapter help to evaluate the respective efficiency of various mesh-based approximation strategies for converging approximations to a non-smooth function. Mesh refinement by local division, like the isotropic AMR, is a classical one. It can be compared with various mesh adaptation strategies, using an isotropic mesh regeneration or anisotropic one. All these methods are evaluated and compared from their performance in the approximation of discontinuous, piecewise regular functions. This is based on theoretical remarks and counterexamples. It is completed by a few concrete numerical calculations. The output of the study is summed up as follows:

– Uniform refinement is definitively an inefficient approach for discontinuous functions.

– Isotropic adaptative strategies show in 2D and in 3D cases an *asymptotical accuracy order limited by barriers*, yet better than first order (order $3/4$ for 3D) when they are applied to discontinuities beared by curves in 2D (surfaces in 3D).

This remark explains in particular the disappointing results obtained when isotropic refinement is applied to the capturing of shock surfaces around 3D aircrafts. In the elliptic case of computational structural mechanics, discontinuities in the medium properties cause only derivative discontinuities. However, their computation with an isotropic adaptative strategy cannot be better than of order $3/2$ in 3D. In order to (try to) escape these limitations, it seems essential to apply anisotropic mesh adaptive methods, especially in the 3D case. Several anisotropic mesh adaptation methods are proposed in the literature and two good examples can be found in Fortin et al. (1999) and Castro-Díaz et al. (1996). They allow a much weaker refinement in the directions, which are tangent to the singularity. Then the above barrier does not apply and it is very probable, but this is yet almost a conjecture, that, for plane singularities, any order of accuracy can be obtained with good anisotropic mesh adaptative strategies.

To summarize, obtaining a high order of convergence toward a function with a local singularity will depend of the following three factors:

– the accuracy order α of the interpolation or approximation;

– the choice of an adequate sensor concentrating the mesh in an anisotropic manner, with a strength adequately tuned with α (stronger with a higher α);

– the ability of the mesh regenerator to follow the indication of the sensor, and in particular to apply mesh density with very sudden and strong variations.

Second, we consider *highly heterogeneous smooth functions*. These functions are smooth, but they involve details in their variations, which are much smaller than expected, in such a way that quasi-uniform discretization approaches produce an insufficient capturing of the details and a practical numerical convergence very similar to the convergence of non-smooth functions, that is, *first-order convergence* in the conditions of first-order convergence identified by the barriers of singular case. For meshes sufficiently fine, convergence at truncation order is theoretically possible but practically not affordable. Such an event would take the appearance of a long phase of first-order convergence followed finally, if we are lucky, by a *late higher-order convergence*. Further, this view is optimistic since difference sizes of details may produce alternations of first-order and higher order convergence! In contrast, adequate anisotropic mesh adaptation methods will give access to the *early capturing effect*, that is, the starting of the higher order convergence phase when mesh size has reached a very affordable number of degrees of freedom.

Concerning mesh adaptation for unsteady flows, we have put in evidence that adapting the spatial mesh at each time step indeed improves the order of convergence but does not allows to recover the truncation order. A cure for escaping this barrier can be a multi-rate approach, as discussed in Chapter 3 of Volume 2.

2.11. Notes

This chapter uses results from Coudière et al. (2002).

The presence of a singularity keeps a high-order approximation with uniform refinement from producing a convergence at order better than, typically, unity. A cure consists of adding to the smooth basis functions a sufficient collection of singular basis functions. As an example, we refer to the computation at high order of the well-known lid-driven cavity flow in Botella and Peyret (1998). This is very useful for obtaining references solutions, but seems not easily applicable to routine industrial simulations.

3

Mesh Representation

This chapter defines the main notion of this book, the *Riemannian metric* or *continuous mesh* representation of a mesh, and reviews the important properties of metrics in relation with simplicial meshes (triangulations, tetrahedrizations) and P^1 finite-element representation/interpolation, with a special emphasis on anisotropic metrics corresponding to stretched meshes. Operations on metrics and relations with geometric quantities are addressed.

3.1. Introduction

Designing a mesh adaptation strategy for best convergence is a difficult question. We have seen in Chapter 2 that the set of possible meshes in which the adapted mesh is sought is of paramount importance. In the case, where the mesh is parameterized by deformation of a reference one, many works in the literature propose to define the set of mesh coordinates as the solution of a particular system coupled with the discrete partial differential equation (PDE) under study (see, for example, Gnoffo 1983; Cabello et al. 1991; Palmerio 1996). In that case, the fixed mesh topology, initially prescribed by the user, may be not adequate for an optimal adaptation. Conversely, in the case where the user does not wish to fix the topology, but instead wants the algorithm to find it, then the definition of a system the solution of which is the adapted mesh is more difficult. We first observe that this is an ill-posed problem since two meshes can have very different topologies and give about the same local accuracy. Second, it is difficult to find an optimal mesh if we have to investigate inside a set of meshes described by integers and booleans.

These remarks have motivated researchers to represent meshes by continuous functions describing the mesh. These functions can be, for example, the (scalar) local mesh density over the computational domain. From its knowledge and assuming a bounded stretching, it is possible to derive an upper bound for the local truncation error. But this scalar upper bound does not give a perfect idea of the local error since it is not expressed as a function of the local stretching effects.

Two decades ago, several publications (see, for example, Borouchaki and George 1996; Frey and George 1999; Habashi et al. 2000) proposed to model the local stretching by means of a non-scalar field, the *Riemannian metric*. An adapted metric is then specified by an argument of equidistribution of an interpolation error related to the PDE solution.

In this chapter, we formalize this notion into a general continuous framework, referred in the sequel as *Riemannian metric*, or *continuous mesh*.

3.2. An introductory example

Let $\Omega = [-1, 1] \times [-1, 1]$ be a square domain of $\mathbb{R}^2$ and $\mathcal{H}_1$ be a uniform mesh of Ω containing 144 vertices (121 triangles). $\mathcal{H}_1$ is depicted in Figure 3.1 (left). We consider the function:

$$f: \mathbb{R}^2 \longrightarrow \mathbb{R} \quad (x, y) \longmapsto f(x, y) = \sqrt{1 - x^2},$$

representing a half-cylinder (see Figure 3.1, middle).

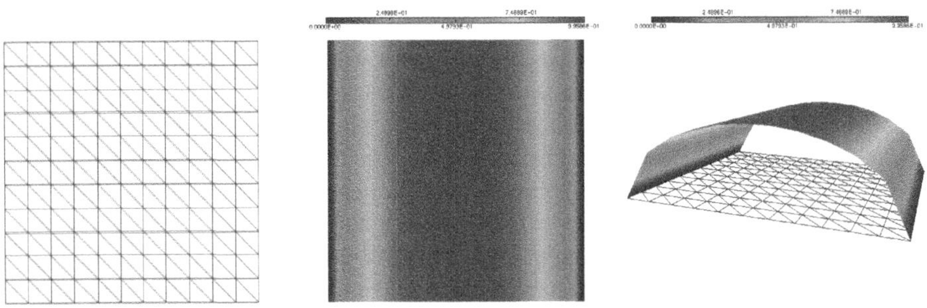

$\mathcal{H}_1$ $\qquad$ f on $\mathcal{H}_1$ $\qquad$ $\Pi_{\mathcal{H}_1} f$

Figure 3.1. *Left, uniform mesh $\mathcal{H}_1$. Middle, iso-values of function f on $\mathcal{H}_1$. Right, representation of $\Pi_{\mathcal{H}_1} f$. For a color version of this figure, see www.iste.co.uk/dervieux/meshadaptation1*

DEFINITION 3.1.– *We call the linear interpolate of f the function $\Pi_{\mathcal{H}_1} f$ that is equal to f at each vertex of $\mathcal{H}_1$ and that is linear inside each triangle.*

$\Pi_{\mathcal{H}_1} f$ is represented in Figure 3.1 (right).

DEFINITION 3.2.– *The interpolation error* $e(f)$ *of* f *on* $\mathcal{H}_1$ *is the gap* $\Pi_{\mathcal{H}_1} f - f$ *between* f *and its linear interpolate.*

We can quantify in L^p norm this error:

$$e(f) = \|f - \Pi_{\mathcal{H}_1} f\|_{\mathbf{L}^p(\Omega)}.$$

For this example, the interpolation error in several norms is given in Table 3.1.

Norm	$\mathbf{L}^1$	$\mathbf{L}^2$	$\mathbf{L}^\infty$
$\|f - \Pi_{\mathcal{H}_1} f\|_{\mathbf{L}^p(\Omega)}$	0.029	0.059	0.133
$\|f - \Pi_{\mathcal{H}_2} f\|_{\mathbf{L}^p(\Omega)}$	0.008	0.005	0.014

Table 3.1. *Interpolation error of function f on uniform mesh $\mathcal{H}_1$ (top) and adapted mesh $\mathcal{H}_2$ (bottom) for several norms*

In this context, the problem of mesh adaptation can be formulated as follows:

How to reduce the interpolation error with the same number of vertices?

This is mathematically written as follows:

$$\text{Find } \mathcal{H}, |\mathcal{H}| = 144 \quad \text{and} \quad \mathcal{H} = \text{Argmin } \|f - \Pi_{\mathcal{H}} f\|_{\mathbf{L}^p(\Omega)}.$$

If the properties of function f are geometrically analyzed, we note that

1) the largest variation of the function (curvature) is in direction x;

2) there is no variation of the function in direction y, that is, the function does not depend on y.

We call this function *anisotropic* as it behaves differently in the x and y directions. In comparison, the Gaussian function $g(x, y) = \exp(-30\,(x^2 + y^2))$ is isotropic because it behaves (varies) similarly in all directions (see Figure 3.2).

To improve the function representation, we propose to define a new mesh using the anisotropic properties of f. Indeed, it is intuitively clear that adding vertices in the y-direction is useless for improving the representation of f while refining the x-axis will provide a more accurate representation. This intuition is illustrated in Figure 3.3 (right). Consequently, we propose to build the mesh having 144 vertices, which is the most accurate in the x-direction and the less accurate in the y-direction meaning that we put uniformly all the vertices on lines $y = -1$ and $y = 1$. This mesh is depicted in Figure 3.4 (left) and is a typical anisotropic or stretched mesh. Visually, $\Pi_{\mathcal{H}_2} f$ clearly seems more accurate than $\Pi_{\mathcal{H}_1} f$. This fact is confirmed by analyzing the interpolation error in several norms (see Table 3.1). We notice that the interpolation error has been divided by 10 in $\mathbf{L}^2$ and $\mathbf{L}^\infty$ norms, and by almost 4 in $\mathbf{L}^1$ norm!

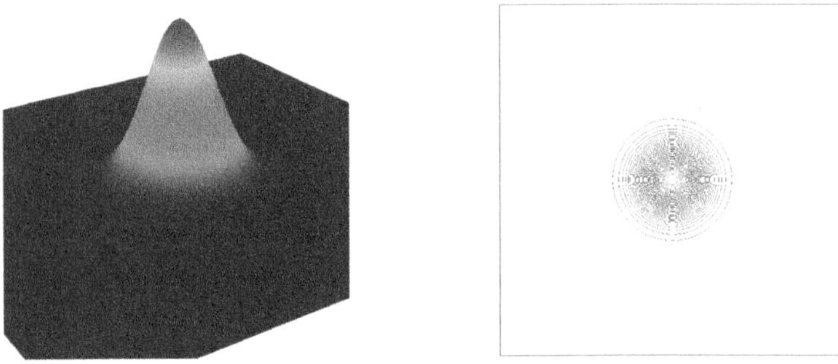

Figure 3.2. *Three-dimensional representation and iso-lines of Gaussian function g. For a color version of this figure, see www.iste.co.uk/dervieux/meshadaptation1*

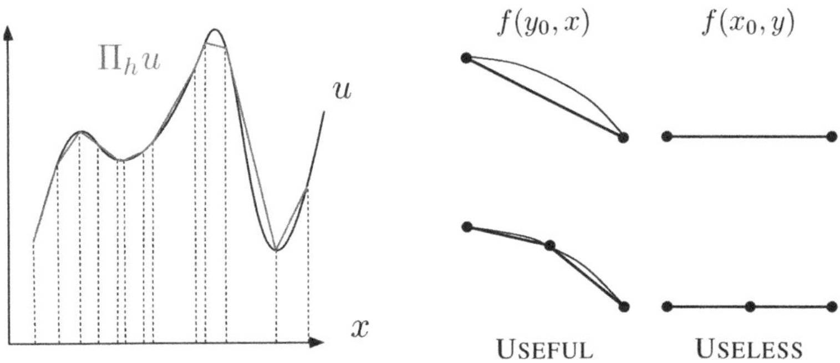

Figure 3.3. *Left, illustration of the linear interpolate $\Pi_h u$ of a function u. Right, refining the mesh in the x-direction improves the representation of function f while adding vertices in the y-direction is useless. For a color version of this figure, see www.iste.co.uk/dervieux/meshadaptation1*

In this simple example, we have illustrated that generating manually a non-uniform mesh using the anisotropy property of function f allows its representation to be significantly improved. We have manually adapted the mesh.

Anisotropic mesh adaptation consists of setting up this process *automatically*. However, we must first answer the following questions:

1) *How do we communicate with an adaptive mesh generator?*

2) *How do we measure or quantify the desired mesh size and anisotropy from a given function?*

In this chapter, we will show that the fundamental concept of *metric* is an elegant and very efficient way to answer these questions through metric-based mesh adaptation.

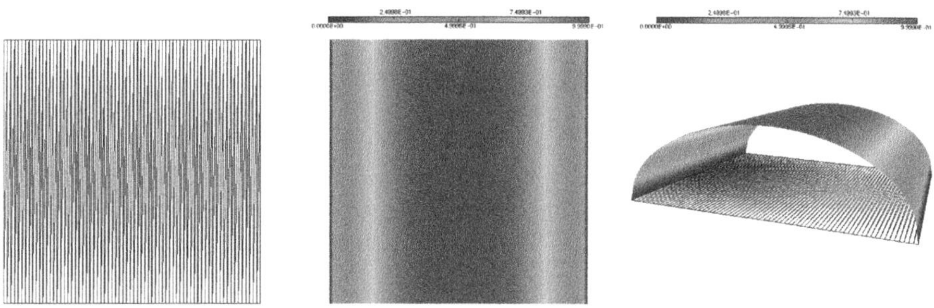

$\mathcal{H}_2$ $\qquad$ f on $\mathcal{H}_2$ $\qquad$ $\Pi_{\mathcal{H}_2} f$

Figure 3.4. *Left, "adapted" mesh $\mathcal{H}_2$. Middle, iso-values of function f on $\mathcal{H}_2$. Right, representation of $\Pi_{\mathcal{H}_2} f$. For a color version of this figure, see www.iste.co.uk/dervieux/meshadaptation1*

3.3. Euclidean metric space

In order to be self-contained, we recall the differential geometry notions that are used in the sequel. It mainly concerns the computation of lengths in different kinds of spaces: the Euclidean space, Euclidean metric spaces and Riemannian metric spaces. A complete review and the mathematical study of these spaces are available in Berger (1987a,b, 2003) and do Carmo (1976).

In the sequel, bold face symbols, as $\mathbf{a}, \mathbf{b}, \mathbf{u}, \mathbf{v}, \mathbf{x}, \ldots$, denote vectors or points of $\mathbb{R}^3$. Vector coordinates are denoted by $\mathbf{x} = (x_1, x_2, x_3)$. The natural dot product between two vectors $\mathbf{u}$ and $\mathbf{v}$ of $\mathbb{R}^3$ is

$$\langle \mathbf{u}, \mathbf{v} \rangle = \sum_{i=1}^{3} u_i \, v_i .$$

A *Euclidean metric space* $(\mathbb{R}^3, \mathcal{M})$ is a finite vector space where the dot product is defined by means of a symmetric definite positive tensor $\mathcal{M}$:

$$\langle \mathbf{u}, \mathbf{v} \rangle_{\mathcal{M}} = \langle \mathbf{u}, \mathcal{M}\mathbf{v} \rangle = {}^t\mathbf{u}\mathcal{M}\mathbf{v}, \quad \text{for } (\mathbf{u}, \mathbf{v}) \in \mathbb{R}^3 \times \mathbb{R}^3.$$

The form $\mathcal{M}$ is usually written as a 3×3 matrix that is

i) (symmetric) $\forall (\mathbf{u}, \mathbf{v}) \in \mathbb{R}^3 \times \mathbb{R}^3, \quad \langle \mathbf{u}, \mathcal{M}\mathbf{v} \rangle = \langle \mathbf{v}, \mathcal{M}\mathbf{u} \rangle$;

ii) (positive) $\forall \mathbf{u} \in \mathbb{R}^3, \quad \langle \mathbf{u}, \mathcal{M}\mathbf{u} \rangle \geq 0$;

iii) (definite) $\langle \mathbf{u}, \mathcal{M}\mathbf{u} \rangle = 0 \Longrightarrow \mathbf{u} = \mathbf{0}$.

These properties ensure that $\mathcal{M}$ defines a dot product. In the following, the matrix $\mathcal{M}$ is simply called *a metric tensor* or *a metric*.

EXAMPLE 3.1.– The most familiar example is used by the elementary Euclidean geometry: the three-dimensional *Euclidean space*, which is defined by the identity matrix

$$\mathcal{M} = \mathcal{I}_3 = \begin{pmatrix} 1 & 0 & 0 \\ 0 & 1 & 0 \\ 0 & 0 & 1 \end{pmatrix}. \square$$

The dot product defined by $\mathcal{M}$ makes $\mathbb{R}^3$ become a normed vector space $(\mathbb{R}^3, \|.\|_{\mathcal{M}})$ and a metric vector space $(\mathbb{R}^3, d_{\mathcal{M}}(.,.))$ supplied by the following *norm* and *distance* definitions:

– $\forall \mathbf{u} \in \mathbb{R}^3, \quad \|\mathbf{u}\|_{\mathcal{M}} = \sqrt{\langle \mathbf{u}, \mathcal{M}\mathbf{u} \rangle}$

– $\forall (\mathbf{a}, \mathbf{b}) \in \mathbb{R}^3 \times \mathbb{R}^3, \quad d_{\mathcal{M}}(\mathbf{a}, \mathbf{b}) = \|\mathbf{b} - \mathbf{a}\|_{\mathcal{M}} = \|\mathbf{ab}\|_{\mathcal{M}}$.

In these spaces, the *length* $\ell_{\mathcal{M}}$ of a segment $\mathbf{ab} = [\mathbf{a}, \mathbf{b}]$ is given by the distance between its extremities:

$$\ell_{\mathcal{M}}(\mathbf{ab}) = d_{\mathcal{M}}(\mathbf{a}, \mathbf{b}) = \|\mathbf{ab}\|_{\mathcal{M}}. \qquad [3.1]$$

Note that this property will be generally wrong for a general Riemannian metric space as defined hereafter. In an Euclidean metric space, volumes and angles are well defined. These features are of main interest when dealing with meshing. Given a bounded subset K of $\mathbb{R}^3$, the *volume* of K computed with respect to metric tensor $\mathcal{M}$ is as follows:

$$|K|_{\mathcal{M}} = \sqrt{\det \mathcal{M}}\,|K|_{\mathcal{I}_3}, \qquad [3.2]$$

where $|K|_{\mathcal{I}_3}$ is the Euclidean volume of K. The *angle* between two non-zero vectors $\mathbf{u}$ and $\mathbf{v}$ is defined by the unique real-value $\theta_{\mathcal{M}} \in [0, \pi]$ verifying

$$\cos(\theta_{\mathcal{M}}) = \frac{\langle \mathbf{u}, \mathbf{v} \rangle_{\mathcal{M}}}{\|\mathbf{u}\|_{\mathcal{M}} \|\mathbf{v}\|_{\mathcal{M}}}. \qquad [3.3]$$

In three dimensions, *dihedral angles*[1] of a tetrahedron can be computed in a given Euclidean space from this definition as it is the angle between the faces normals.

We are also able to compute cross product with respect to metric tensor $\mathcal{M}$:

$$\forall (\mathbf{a}, \mathbf{b}) \in \mathbb{R}^3 \times \mathbb{R}^3, \quad \mathcal{M}\mathbf{a} \wedge \mathcal{M}\mathbf{b} = (\det \mathcal{M})\ \mathcal{M}^{-\frac{1}{2}}\ (\mathbf{a} \wedge \mathbf{b})\,.$$

This is very useful in 3D for computing the area of faces with respect to $\mathcal{M}$. Let F be a triangle in $\mathbb{R}^3$:

$$\begin{aligned} |F|_{\mathcal{M}} &= \|\mathcal{M}\mathbf{a} \wedge \mathcal{M}\mathbf{b}\| = \|\mathbf{a}\|_{\mathcal{M}}\, \|\mathbf{b}\|_{\mathcal{M}}\, \sin(\theta_{\mathcal{M}}) \\ \text{or} \quad |F|^2_{\mathcal{M}} &= \|\mathcal{M}\mathbf{a} \wedge \mathcal{M}\mathbf{b}\|^2 = \|\mathbf{a}\|^2_{\mathcal{M}}\, \|\mathbf{b}\|^2_{\mathcal{M}} - \langle \mathbf{a}, \mathcal{M}\mathbf{b} \rangle^2. \end{aligned}$$

Finally, as metric tensor $\mathcal{M}$ is symmetric, it is diagonalizable. It thus admits the following spectral decomposition:

$$\mathcal{M} = \mathcal{R}\,\Lambda\,{}^t\mathcal{R},$$

where $\mathcal{R}$ is an orthonormal matrix composed of the *eigenvectors* $(\mathbf{v}_i)_{i=1,3}$ of $\mathcal{M}$:

$$\mathcal{R} = (\mathbf{v}_1\ \ \mathbf{v}_2\ \ \mathbf{v}_3)\,,$$

verifying ${}^t\mathcal{R}\mathcal{R} = \mathcal{R}{}^t\mathcal{R} = \mathcal{I}_3$. $\Lambda = diag(\lambda_i)$ is a diagonal matrix composed of the *eigenvalues* of $\mathcal{M}$, denoted $(\lambda_i)_{i=1,3}$, which are strictly positive.

1 The dihedral angle is the angle between two intersecting planes, for example, two triangular faces of a tetrahedron.

3.3.1. *Geometric interpretation*

We will often refer to the geometric interpretation of a metric tensor. In the vicinity $\mathcal{V}(\mathbf{a})$ of point $\mathbf{a}$, the set of points that are at distance ε of $\mathbf{a}$ is given by

$$\Phi_{\mathcal{M}}(\varepsilon) = \left\{\mathbf{x} \in \mathcal{V}(\mathbf{a}) \mid {}^t(\mathbf{x}-\mathbf{a})\,\mathcal{M}\,(\mathbf{x}-\mathbf{a}) = \varepsilon^2\right\}.$$

It is sufficient to describe $\Phi_{\mathcal{M}}(1)$ as $\Phi_{\mathcal{M}}(\varepsilon)$ can be deduced from $\Phi_{\mathcal{M}}(1)$ for all ε by homogeneity, $\Phi_{\mathcal{M}}(\varepsilon) = \varepsilon\Phi_{\mathcal{M}}(1)$. Another description of $\Phi_{\mathcal{M}}(1)$ can be given using the spectral decomposition $\mathcal{M} = \mathcal{R}\,\Lambda\,{}^t\mathcal{R}$. In the eigenvectors frame, the initial quadratic form ${}^t(\mathbf{x}-\mathbf{a})\,\mathcal{M}\,(\mathbf{x}-\mathbf{a})$ becomes ${}^t(\tilde{\mathbf{x}}-\tilde{\mathbf{a}})\,\Lambda\,(\tilde{\mathbf{x}}-\tilde{\mathbf{a}})$. Consequently, rewriting $\Phi_{\mathcal{M}}(1)$ in this basis leads to the following definition:

DEFINITION 3.3 (Unit ball).– *The unit ball of $\mathcal{M}$ at point $\mathbf{x}$ is defined by*

$$\begin{aligned} \Phi_{\mathcal{M}}(1) &= \left\{\tilde{\mathbf{x}} \in \mathcal{V}(\tilde{\mathbf{a}}) \mid \sum_{i=1}^{3} \lambda_i\,(\tilde{x}_i - \tilde{a}_i)^2 = 1\right\} \\ &= \left\{\tilde{\mathbf{x}} \in \mathcal{V}(\tilde{\mathbf{a}}) \mid \sum_{i=1}^{3} \left(\frac{\tilde{x}_i - \tilde{a}_i}{h_i}\right)^2 = 1\right\}. \end{aligned} \qquad [3.4]$$

The last relation defines an ellipsoid centered at point $\mathbf{a}$ with its axes aligned with the eigen directions of $\mathcal{M}$. Sizes along these directions are given by $h_i = \lambda_i^{-\frac{1}{2}}$. We denote by $\mathcal{B}_{\mathcal{M}}$ this ellipsoid depicted in Figure 3.5. In the sequel, the set $\mathcal{B}_{\mathcal{M}} = \Phi_{\mathcal{M}}(1)$ is called the *unit ball* of $\mathcal{M}$.

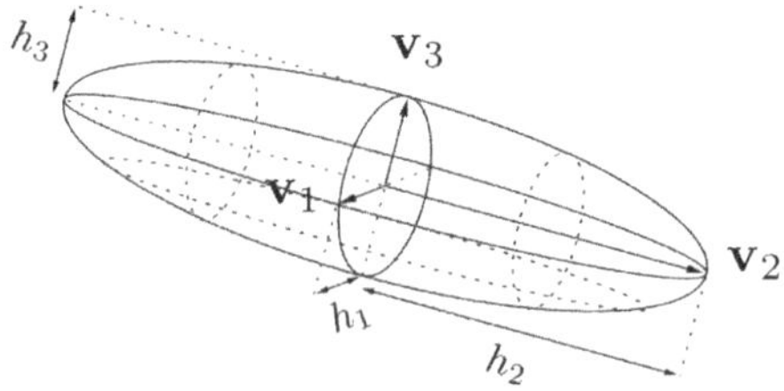

Figure 3.5. *Geometric interpretation of the unit ball $\mathcal{B}_{\mathcal{M}} = \Phi_{\mathcal{M}}(1)$ of $\mathcal{M}$. $\mathbf{v}_i$ are the eigenvectors of $\mathcal{M}$ and $\lambda_i = h_i^{-2}$ are the eigenvalues of $\mathcal{M}$*

3.3.2. Natural metric mapping

The last information handled by a metric tensor $\mathcal{M}$ is the definition of an application that maps the unit ball $\mathcal{B}_{\mathcal{I}_3}$ of identity metric $\mathcal{I}_3$ onto the unit ball $\mathcal{B}_{\mathcal{M}}$ of $\mathcal{M}$. We introduce $\mathcal{M}^{-\frac{1}{2}}$ defined by the spectral decomposition

$$\mathcal{M}^{-\frac{1}{2}} = \mathcal{R}\,\Lambda^{-\frac{1}{2}}\,{}^t\mathcal{R} \quad \text{where} \quad \Lambda^{-\frac{1}{2}} = diag(\lambda_i^{-\frac{1}{2}}) = \begin{pmatrix} \lambda_1^{-\frac{1}{2}} & 0 & 0 \\ 0 & \lambda_2^{-\frac{1}{2}} & 0 \\ 0 & 0 & \lambda_3^{-\frac{1}{2}} \end{pmatrix}.$$

This decomposition exists because $\mathcal{M}$ is definite positive. The mapping

$$\mathcal{M}^{-\frac{1}{2}} : \mathbb{R}^3 \longmapsto \mathbb{R}^3 \quad \mathbf{x} \longrightarrow \mathcal{M}^{-\frac{1}{2}}\,\mathbf{x}$$

provides another description of the ellipsoid $\mathcal{B}_{\mathcal{M}}$:

$$\mathcal{B}_{\mathcal{M}} = \left\{\mathcal{M}^{-\frac{1}{2}}\,\mathbf{x} \mid \|\mathbf{x}\|_2 = 1\right\}.$$

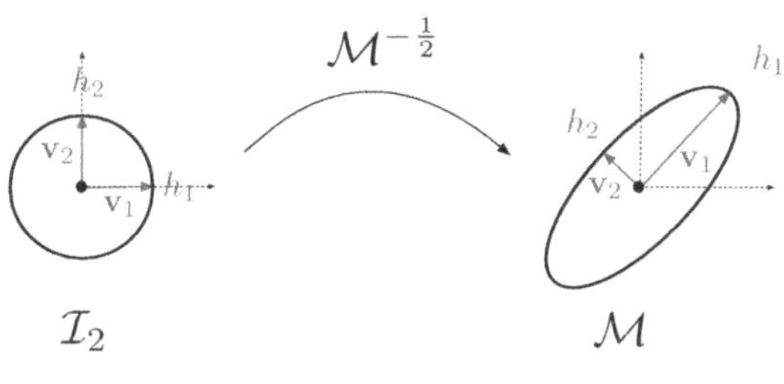

Figure 3.6. *Natural mapping associated with metric $\mathcal{M}$. For a color version of this figure, see www.iste.co.uk/dervieux/meshadaptation1*

3.4. Riemannian metric space

In differential geometry, a Riemannian metric $\mathbf{M}$ on a smooth manifold Ω is a family $\mathbf{x} \mapsto \mathcal{M}(\mathbf{x})$ of smoothly varying inner products applying on the tangent spaces $T_{\mathbf{x}}\Omega$ of Ω. In other words, for each $\mathbf{x} \in \Omega$, $\mathcal{M}(\mathbf{x})$ satisfies

1) $\mathcal{M}(\mathbf{x})(u,v) = \mathcal{M}(\mathbf{x})(v,u)$ for all $u, v \in T_{\mathbf{x}}\Omega$;

2) $\mathcal{M}(\mathbf{x})(u,u) \geq 0$ for all $u \in T_{\mathbf{x}}\Omega$;

3) $\mathcal{M}(\mathbf{x})(u,u) = 0$ if and only if $u = 0$.

Furthermore, $\mathbf{M}$ is smooth in the sense that for any smooth vector fields X and Y, the function $\mathbf{x} \mapsto \mathcal{M}(\mathbf{x})(X(\mathbf{x}), Y(\mathbf{x}))$ is smooth.

In the context of mesh adaptation, the computational domain Ω is in 3D an open subset of $\mathbb{R}^3$, which lies on only one side of its boundary; this boundary being a two-dimensional piecewise smooth manifold. Since Ω is an open space of $\mathbb{R}^3$, the tangent space at any $\mathbf{x} \in \mathbb{R}^3$ is simply $\mathbb{R}^3$. On Ω, we consider a simpler object called a *Riemannian metric space* defined by Ω equipped with $\mathbf{M} = (\mathcal{M}(\mathbf{x}))_{\mathbf{x}\in\Omega}$. This mathematical object can be assimilated to a function that can represents a set of Cartesian surfaces (or graph surfaces). A Cartesian surface $\mathcal{S}$ of $\mathbb{R}^{n+1}$ is a surface (a sub-manifold of $\mathbb{R}^{n+1}$) locally defined by

$$\mathcal{S} = (x_1, \ldots, x_n, \sigma(x_1, \ldots, x_n)) \in \mathbb{R}^{n+1},$$

with σ a $\mathcal{C}^2$ scalar function defined on $\mathbb{R}^n$.

Evaluating geometrical quantities in a Riemannian metric space is equivalent to evaluating these quantities on the underlying Cartesian surfaces.

But, we do not need to know such surfaces. In a rough outline, a Riemannian metric space curves the parameterization space, that is, the computational domain. Let us give an example to illustrate Riemannian metric spaces.

EXAMPLE 3.2.– This example provides a practical visualization of a Riemannian metric space. Moreover, it illustrates why a Riemannian structure curves the space. Let $\mathcal{S} \subset \mathbb{R}^3$ be a Cartesian surface defined by

$$\mathcal{S} = \left\{(x, y, z) \in \mathbb{R}^3 \,|\, z = \sigma(x, y) = x^2 - y^2 \text{ and } (x, y) \in \Omega\right\},$$

where $\Omega = \left\{(x, y) \in \mathbb{R}^2 \,|\, 1 \leq x, y \leq 1\right\}$. This Cartesian surface is represented in Figure 3.7 (left). We consider a curve γ drawn in Ω and parameterized by $t \in \mathbb{R}$:

$$\gamma : [0, 1] \longrightarrow \Omega \subset \mathbb{R}^2 \quad t \longmapsto \gamma(t) = (x(t), y(t)).$$

Its image $\mathbf{c}$ on Cartesian surface $\mathcal{S}$ is a curve of $\mathbb{R}^3$:

$$\begin{aligned} \mathbf{c} : [0, 1] &\longrightarrow \mathcal{S} \subset \mathbb{R}^3 \\ t &\longmapsto \mathbf{c}(t) = (\gamma(t), \sigma(\gamma(t))) = (x(t), y(t), \sigma(x(t), y(t))). \end{aligned}$$

The length of curve $\mathbf{c}$ is then given by

$$\ell(\mathbf{c}) = \int_0^1 \left\| \frac{\mathrm{d}\mathbf{c}}{\mathrm{d}t}(t) \right\| \mathrm{d}t.$$

From the chain rule, we get

$$\left\| \frac{\mathrm{d}\mathbf{c}}{\mathrm{d}t}(t) \right\|^2 = \left\| \frac{\mathrm{d}\gamma}{\mathrm{d}t}(t) \right\|^2 + \left(\mathrm{d}\sigma_{\gamma(t)} \left[\frac{\mathrm{d}\gamma}{\mathrm{d}t}(t) \right] \right)^2.$$

For $\mathcal{S}$ and by hiding the dependence on t, it yields

$$\begin{aligned} \|d\mathbf{c}\|^2 &= dx^2 + dy^2 + dz^2 = dx^2 + dy^2 + (d(x^2 - y^2))^2 \\ &= dx^2 + dy^2 + (2xdx - 2ydy)^2 \\ &= dx^2 + dy^2 + 4x^2dx^2 - 8xydxdy + 4y^2dy^2 \\ &= (dx\ dy)\mathcal{M}(x,y)^t(dx\ dy), \end{aligned}$$

where

$$\mathcal{M}(x,y) = \begin{pmatrix} 1 & 0 \\ 0 & 1 \end{pmatrix} + 4 \begin{pmatrix} x^2 & -xy \\ -xy & y^2 \end{pmatrix} = \mathcal{I}_2 + \mathcal{N}.$$

This quadratic form, $\|\mathrm{d}\mathbf{c}\|^2$, is the first fundamental form of Cartesian surface $\mathcal{S}$, which is represented by $\mathcal{M}$ under matrix form. It exhibits the Riemannian metric space associated with this surface. This Riemannian metric space is shown in Figure 3.7 (right) where the unit ball of the metric at some points of the domain has been drawn. By construction, computing distances in $(\mathcal{M}(\mathbf{x}))_{\mathbf{x}\in\Omega}$ is equivalent to compute distances on Cartesian surface $\mathcal{S} \subset \mathbb{R}^3$. In other words, let $\mathbf{a}$ and $\mathbf{b}$ be two points of Ω and $\gamma(t) = \mathbf{a} + t\mathbf{ab}$ be the straight parameterization of the path of edge $\mathbf{ab}$ in Ω. Let $\mathcal{S}\mathbf{a}$ and $\mathcal{S}\mathbf{b}$ be the image of these points on $\mathcal{S}$. Then, $\ell_{\mathcal{M}}(\mathbf{ab})$ is equal to $\ell_{\mathcal{S}}(\mathcal{S}\mathbf{a}\mathcal{S}\mathbf{b})$, which is the length of the curved path on $\mathcal{S}$ between $\mathcal{S}\mathbf{a}$ and $\mathcal{S}\mathbf{b}$ by following the image of $\gamma(t)$ by the mapping. In this example, we have considered the Riemannian metric space inherited from the Cartesian parameterization of $\mathcal{S}$ on Ω. Thus, this surface represents the curvature induced by the Riemannian structure on Ω. However, the reciprocality that consists of exhibiting a manifold, that is, the Cartesian hypersurface corresponding to the curved domain, from a given Riemannian metric space is a difficult task. Solving this problem requires to seek a hypersurface from its first and second fundamental forms (Ciarlet 2005). In our case, we only know the first fundamental form. □

Now let us extend the notions of length, volume and angle defined in section 3.3 for Euclidean metric spaces to Riemannian metric spaces. They will be the main functions used by the mesher in such spaces. Fortunately, these notions can be easily derived in the context of meshing because we are not interested in evaluating these quantities on the Riemannian manifold. Indeed, with regard to edge length computation, we do not want to compute the distance between two points that requires to find the shortest path on the curved manifold between these two points and to compute the length of the geodesic, but to compute the length of the path between these two points defined by the straight line parameterization, that is, the segment representing the edge between these two points in the parameterization space, which is our computational domain. To take into account the variation of the metric along the edge, the edge length is evaluated with an integral formula.

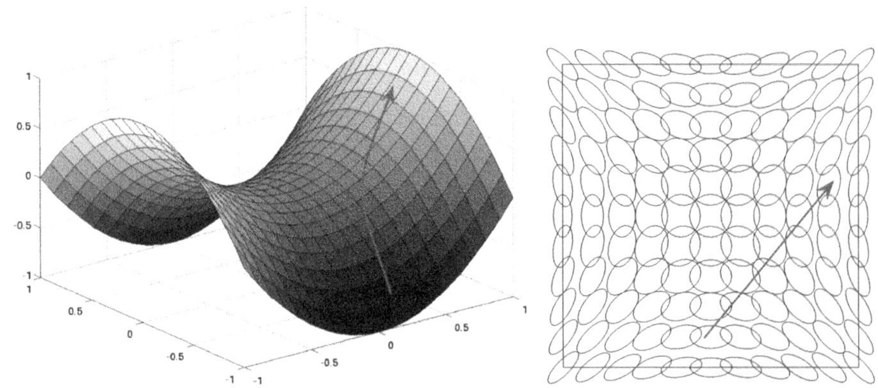

Figure 3.7. *Left: Example of a Cartesian surface embedded in $\mathbb{R}^3$. Right: Geometric visualization of a Riemannian metric space $(\mathcal{M}(\mathbf{x}))_{\mathbf{x}\in[0,1]\times[0,1]}$ associated with this surface. At some points $\mathbf{x}$ of the domain, the unit ball of $\mathcal{M}(\mathbf{x})$ is drawn. For a color version of this figure, see www.iste.co.uk/dervieux/meshadaptation1*

DEFINITION 3.4.– *In Riemannian metric space $\mathbf{M} = (\mathcal{M}(\mathbf{x}))_{\mathbf{x}\in\Omega}$, the length of edge $\mathbf{ab}$ is computed using the straight line parameterization in domain Ω, $\gamma(t) = \mathbf{a} + t\,\mathbf{ab}$, where $t \in [0,1]$:*

$$\ell_{\mathcal{M}}(\mathbf{ab}) = \int_0^1 \|\gamma'(t)\|_{\mathcal{M}}\, \mathrm{d}t = \int_0^1 \sqrt{{}^t\mathbf{ab}\; \mathcal{M}(\mathbf{a} + t\,\mathbf{ab})\; \mathbf{ab}}\, \mathrm{d}t. \qquad [3.5]$$

Figure 3.8 depicts iso-values of segment length from the origin for different Riemannian metric spaces.

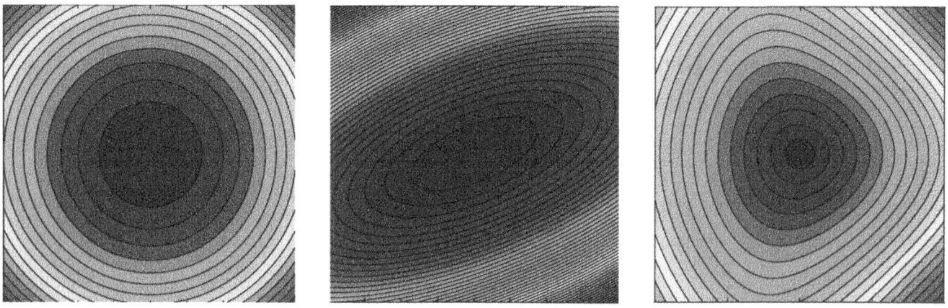

Figure 3.8. *Iso-values of the function $f(\mathbf{x}) = \ell_{\mathcal{M}}(\mathbf{ox})$ where $\mathbf{o}$ is the origin, that is, $f(\mathbf{x})$ is the segment length issued from the origin, for different Riemannian metric spaces. Left: In the canonical Euclidean space $([-1,1]\times[-1,1], I_2)$. Middle: in an Euclidean metric space $([-1,1]\times[-1,1], \mathcal{M})$ with $\mathcal{M}$ constant. Right: in a Riemannian metric space $(\mathcal{M}(\mathbf{x}))_{\mathbf{x}\in[-1,1]^2}$ with a varying metric tensor field. For a color version of this figure, see www.iste.co.uk/dervieux/meshadaptation1*

The plotted function is $f(\mathbf{x}) = \ell_{\mathcal{M}}(\mathbf{ox})$, where $\mathbf{o}$ is the origin of the plane. The iso-values are isotropic for the Euclidean space. They are anisotropic in the case of an Euclidean metric space defined by $\mathcal{M}$. The two principal directions of $\mathcal{M}$ clearly appear. In the case of a Riemannian metric space $(\mathcal{M}(\mathbf{x}))_{\mathbf{x}\in\Omega}$, all previous symmetries are lost.

To conclude this section, let us introduce the notions of volume and angle in Riemannian metric spaces.

DEFINITION 3.5.– *Given a bounded subset K of Ω, the volume of K computed with respect to Riemannian metric space $(\mathcal{M}(\mathbf{x}))_{\mathbf{x}\in\Omega}$ is*

$$|K|_{\mathcal{M}} = \int_K \sqrt{\det \mathcal{M}(\mathbf{x})}\, \mathrm{d}\mathbf{x}. \qquad [3.6]$$

For a Riemannian manifold, the angle between two curves is the angle between their tangent vectors at their point of intersection in the tangent plane. In our particular context, edge and tangent vector coincide, hence, we have the following:

DEFINITION 3.6.– *The angle between two edges $\mathbf{pq}$ and $\mathbf{pr}$ of Ω in Riemannian metric space $(\mathcal{M}(\mathbf{x}))_{\mathbf{x}\in\Omega}$ is defined by the unique real-value $\theta \in [0,\pi]$ verifying*

$$\cos(\theta) = \frac{\langle \mathbf{pq}, \mathbf{pr}\rangle_{\mathcal{M}(\mathbf{p})}}{\|\mathbf{pq}\|_{\mathcal{M}(\mathbf{p})}\|\mathbf{pr}\|_{\mathcal{M}(\mathbf{p})}}. \qquad [3.7]$$

In three dimensions, the computation of a dihedral angle in a Riemannian metric space requires to use an integral formula as this angle varies depending on the position on the edge sharing both faces.

DEFINITION 3.7.– *The dihedral angle between two faces* $[\mathbf{pqr}]$ *and* $[\mathbf{sqp}]$ *of* Ω *sharing* $\mathbf{pq}$ *as common edge in Riemannian metric space* $(\mathcal{M}(\mathbf{x}))_{\mathbf{x}\in\Omega}$ *is defined by the real-value* $\theta \in [0, \pi]$ *verifying*

$$\cos(\theta) = \int_0^1 \frac{\langle \mathbf{n_{pqr}}, \mathbf{n_{sqp}} \rangle_{\mathcal{M}(\mathbf{p}+t\mathbf{pq})}}{\|\mathbf{n_{pqr}}\|_{\mathcal{M}(\mathbf{p}+t\mathbf{pq})} \|\mathbf{n_{sqp}}\|_{\mathcal{M}(\mathbf{p}+t\mathbf{pq})}} \, dt, \qquad [3.8]$$

where $\mathbf{p} + t\,\mathbf{pq}$ *with* $t \in [0, 1]$ *is the straight line parameterization of edge* $\mathbf{pq}$ *in domain* Ω *and,* $\mathbf{n_{pqr}}$ *and* $\mathbf{n_{sqp}}$ *are the oriented normals to* $[\mathbf{pqr}]$ *and* $[\mathbf{sqp}]$*, respectively.*

3.5. Generation of adapted anisotropic meshes

To generate anisotropic meshes, one must be able to prescribe at each point of the domain a privileged size and orientation of the elements. This information will be transmitted to the mesher, which will try to best fit these demands. The use of Riemannian metric spaces is an elegant and efficient way to achieve this goal. Indeed, it is possible for a mesher to work in a Riemannian metric space as the notions of length, volume and angle are well-posed. The main idea of metric-based mesh adaptation, initially introduced in George et al. (1991), is to generate a *unit mesh* in a prescribed Riemannian metric space. In this section, we first introduce the definition of unit element and then the notion of unit mesh.

3.5.1. *Unit element*

DEFINITION 3.8.– *An element* K *is unit with respect to a metric* $\mathcal{M}$ *if the length of all its edges is unit in the metric* $\mathcal{M}$.

For instance, a tetrahedron K defined by its list of edges $(\mathbf{e_i})_{i=1..6}$ is unit with respect to $\mathcal{M}$ if

$$\forall i = 1, ..., 6, \ \ \ell_{\mathcal{M}}(\mathbf{e_i}) = 1,$$

and its volume is $|K|_{\mathcal{M}} = \dfrac{\sqrt{2}}{12}$. Figure 3.9 gives two examples of unit elements for two different metric tensors.

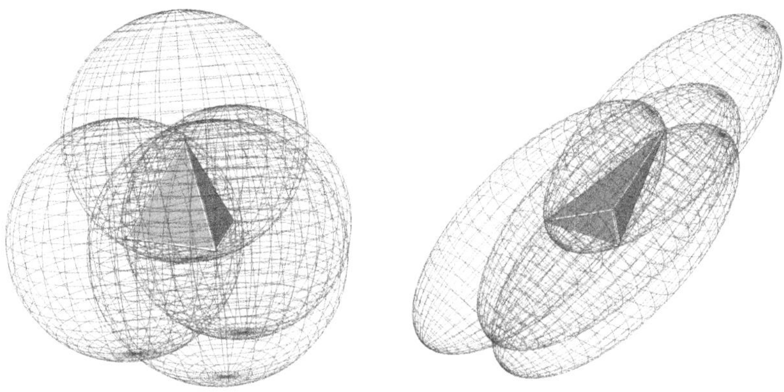

Figure 3.9. *3D examples of a unit element with respect to $\mathcal{I}_3$ (left) and to an anisotropic metric tensor $\mathcal{M}$ (right). In each case, the unit ball of the metric is drawn at each vertex of the unit element. For a color version of this figure, see www.iste.co.uk/dervieux/meshadaptation1*

Let $\mathcal{M}$ be a metric tensor, there exists a non-empty infinite set of unit elements with respect to $\mathcal{M}$, as illustrated in Figure 3.10. Conversely, given an element K such that $|K|_{\mathcal{I}_n} \neq 0$, there is a unique metric tensor $\mathcal{M}$ for which element K is unit with respect to $\mathcal{M}$.

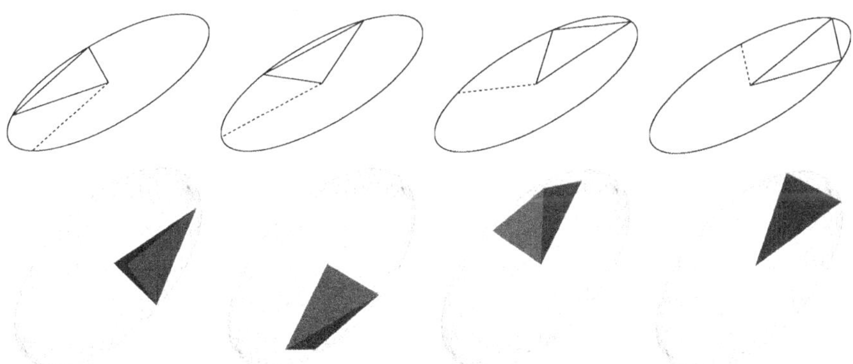

Figure 3.10. *Several unit elements with respect to a metric tensor in 2D and 3D. For a color version of this figure, see www.iste.co.uk/dervieux/meshadaptation1*

3.5.2. *Geometric invariants*

So far, definition 3.8 has established relationships between unit elements and the relative metric tensor. Other properties exist. They connect the geometric properties of unit elements to the linear algebra properties of metric tensors. The following proposition gives geometric invariants that hold for all unit elements with respect to a metric tensor.

PROPOSITION 3.1 (geometric invariants).– Let $\mathcal{M}$ be a metric tensor and K be a unit element with respect to $\mathcal{M}$. We denote by $(\mathbf{e}_i)_{i=1,6}$ its edges list (see conventions in Figure 3.11) and by $|K|$ its Euclidean volume. Then the following invariants hold:

– standard invariants,

$$\forall\,(\mathbf{e}_i,\mathbf{e}_j),\qquad \begin{cases} {}^t\mathbf{e}_i\,\mathcal{M}\,\mathbf{e}_i = 1, \\ 2\,{}^t\mathbf{e}_i\,\mathcal{M}\,\mathbf{e}_j + 1 = 0 \text{ if } i \neq j. \end{cases} \qquad [3.9]$$

– invariant related to the Euclidean volume $|K|$,

$$\boxed{|K| = \frac{\sqrt{3}}{4}\,\det(\mathcal{M}^{-\frac{1}{2}}) \text{ in 2D} \quad\text{and}\quad |K| = \frac{\sqrt{2}}{12}\,\det(\mathcal{M}^{-\frac{1}{2}}) \text{ in 3D.}} \qquad [3.10]$$

– invariant related to the square length of the edges for all symmetric matrix H,

$$\boxed{\begin{aligned} \sum_{i=1}^{3} {}^t\mathbf{e}_i\,H\mathbf{e}_i &= \frac{3}{2}\text{trace}(\mathcal{M}^{-\frac{1}{2}}H\mathcal{M}^{-\frac{1}{2}}) \text{ in 2D,} \\ \sum_{i=1}^{6} {}^t\mathbf{e}_i\,H\mathbf{e}_i &= 2\,\text{trace}(\mathcal{M}^{-\frac{1}{2}}H\mathcal{M}^{-\frac{1}{2}}) \text{ in 3D.} \end{aligned}} \qquad [3.11]$$

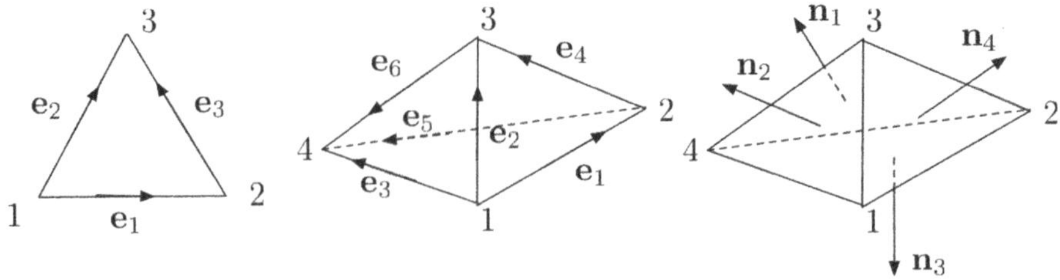

Figure 3.11. *Conventions used to enumerate the edges and the faces of a triangle and a tetrahedron*

PROOF.– The first invariant of relation [3.9] stems from the definition of a unit element. The second invariant of relation [3.9] states that the angle between two edges of a unit element face is constant in the metric. Let $(\mathbf{e}_i, \mathbf{e}_j)$ be a couple of edges of element K; this couple defines a face. We denote by $\mathbf{e}_k$ the third edge of this face. According to the conventions depicted in Figure 3.11, these edges verify that $\mathbf{e}_i + \mathbf{e}_j - \mathbf{e}_k = 0$. Expanding the relation

$$^t(\mathbf{e}_i + \mathbf{e}_j + \mathbf{e}_k)\, \mathcal{M}\, (\mathbf{e}_i + \mathbf{e}_j - \mathbf{e}_k) = 0$$

leads to the second invariant of relation [3.9].

Invariant [3.10] is proved by a direct integration. Given a unit element K for $\mathcal{M}$, there exists a unique regular tetrahedron K_0, which is unit with respect to I_3, such that $K = \mathcal{M}^{-\frac{1}{2}} K_0$. The volume of K is then given by

$$|K| = \int_K 1\, \mathrm{d}\mathbf{x} = \int_{K_0} \det(\mathcal{M}^{-\frac{1}{2}})\, \mathrm{d}\mathbf{x} = \det(\mathcal{M}^{-\frac{1}{2}})\, |K_0|,$$

where $|K_0| = \frac{\sqrt{2}}{12}$. The same proof applies in 2D.

Invariant [3.11] is first proved in the simpler case where $H = I_3$ and $\mathcal{M} = I_3$. The general case will be deduced from this proof. Let us consider the regular tetrahedron $K_0 = (\mathbf{x_1}, \mathbf{x_2}, \mathbf{x_3}, \mathbf{x_4})$ unit for I_3 defined by the list of vertices

$$\mathbf{x_1} = (0,0,0)\,,\ \mathbf{x_2} = (1,0,0)\,,\ \mathbf{x_3} = \left(\frac{1}{2}, \frac{\sqrt{3}}{2}, 0\right)\,,\ \text{and } \mathbf{x_4} = \left(\frac{1}{2}, \frac{\sqrt{3}}{6}, \sqrt{\frac{2}{3}}\right).$$

The proof does not depend on this specific choice of coordinates. We first demonstrate the following preliminary result: *For all lines (D) passing through one of the vertices of K_0, the sum of the square lengths of the edges projected on (D) is invariant.* Without loss of generality, we assume that (D) passes through the vertex $\mathbf{x_1}$ of K_0. If (D) is defined by the vector

$$\mathbf{n} = (\cos(u)\cos(v), \cos(u)\sin(v), \sin(u)),$$

with $(u,v) \in \mathbb{R}^2$, then the length of the first three edges of K_0 projected on (D) is given by

$$a = \mathbf{e_1} \cdot \mathbf{n} = \cos(u)\cos(v), \quad b = \mathbf{e_2} \cdot \mathbf{n} = \frac{1}{2}\cos(u)\cos(v) + \frac{\sqrt{3}}{2}\cos(u)\sin(v),$$
$$c = \mathbf{e_3} \cdot \mathbf{n} = \frac{1}{2}\cos(u)\cos(v) + \frac{\sqrt{3}}{6}\cos(u)\sin(v) + \sqrt{\frac{2}{3}}\sin(u),$$

with conventions in Figure 3.11. A direct trigonometric calculus shows that the sum of the square length of all the edges projected on (D) is equal to 2. Indeed, it comes

$$\begin{aligned}\Sigma &= a^2 + b^2 + c^2 + (b-c)^2 + (c-a)^2 + (a-b)^2 \\ &= 3\,a^2 + 3\,b^2 + 3\,c^2 - 2ab - 2ac - 2bc.\end{aligned}$$

After expanding and factorizing, Σ reads

$$\Sigma = 2\cos(u)^2\cos(v)^2 + 2\cos(u)^2\sin(v)^2 + 2\sin(u)^2 = 2.$$

When $\mathcal{M}$ is different from I_3, we use the mapping $\mathcal{M}^{-\frac{1}{2}}$ that maps the unit ball of I_3 onto the unit ball of $\mathcal{M}$. With regard to line (D), we select the specific line, which has for direction vector one of the eigenvectors of $\mathcal{M}$, for example, $\mathbf{v}_j$, and which is passing through $\mathbf{x_1}$. The lengths a, b and c are thus multiplied by h_j, which is the size prescribed by $\mathcal{M}$ in the direction $\mathbf{v}_j$. Consequently, the square length of the edges projected on (D) are multiplied by h_j^2. It gives

$$\Sigma_j = \sum_{i=1}^{6} ({}^t\mathbf{e}_i\,\mathbf{v}_j)^2 = h_j^2\,\Sigma = 2\,h_j^2.$$

Considering the previous relation for all the principal directions of $\mathcal{M}$ and summing the Σ_j complete the proof:

$$\sum_{i=1}^{6} \|\mathbf{e}_i\|_2^2 = \sum_{j=1}^{3}\sum_{i=1}^{6} ({}^t\mathbf{e}_i\,\mathbf{v}_j)^2 = 2\left(h_1^2 + h_2^2 + h_3^2\right) = 2\,\mathrm{trace}(\mathcal{M}^{-1}). \quad [3.12]$$

We now consider the case where a symmetric matrix H is involved in the estimation. As H is symmetric, it has real eigenvalues $(\mu_i)_{i=1,3}$ along the principal directions $(\mathbf{u}_i)_{i=1,3}$. Let K be a regular tetrahedron given by its list of edges

$(\mathbf{e}_i)_{i=1,6}$. According to the preliminary result, the sum of the projected square length of edges $(\mathbf{e}_i)_{i=1,6}$ on each principal direction $\mathbf{u}_j$ is equal to 2. We deduce

$$\mu_j \sum_{i=1}^{6} ({}^t\mathbf{e}_i\, \mathbf{u}_j)^2 = 2\,\mu_j \quad \text{for } j \in [1,3].$$

Summing the previous equalities leads to

$$\sum_{j=1}^{3} \sum_{i=1}^{6} \mu_j\, ({}^t\mathbf{e}_i\, \mathbf{u}_j)^2 = 2\,(\mu_1 + \mu_2 + \mu_3).$$

The previous equality reads

$$\sum_{i=1}^{6} {}^t\mathbf{e}_i\, H \mathbf{e}_i = 2\,\mathrm{trace}(H).$$

The general case is deduced by taking $\mathcal{M}^{-\frac{1}{2}}\, H \mathcal{M}^{-\frac{1}{2}}$ as symmetric matrix. In that case, each edge $\mathbf{e}_i$ of the regular tetrahedron is mapped on $\tilde{\mathbf{e}}_i = \mathcal{M}^{-\frac{1}{2}}\mathbf{e}_i$. The new tetrahedron defined by edges $(\tilde{\mathbf{e}}_i)_{i=1,6}$ is unit with respect to $\mathcal{M}$. This concludes the proof.□

Other geometric invariants can be found in Loseille (2008).

3.5.3. *Global duality*

When dealing with a Riemannian metric space $\mathbf{M} = (\mathcal{M}(\mathbf{x}))_{\mathbf{x}\in\Omega}$, $\mathbf{M}$ is first rewritten in order to distinguish local properties from global ones. We recall the spectral decomposition of $\mathcal{M}(\mathbf{x})$:

$$\mathbf{M}:\ \mathbf{x} \in \Omega \mapsto \mathcal{M}(\mathbf{x}) = \mathcal{R}(\mathbf{x})\, \Lambda(\mathbf{x})\, {}^t\mathcal{R}(\mathbf{x}),$$

where diagonal matrix $\Lambda(\mathbf{x})$ is

$$\begin{bmatrix} \lambda_1(\mathbf{x}) & & \\ & \lambda_2(\mathbf{x}) & \\ & & \lambda_3(\mathbf{x}) \end{bmatrix} = \begin{bmatrix} h_1^{-2}(\mathbf{x}) & & \\ & h_2^{-2}(\mathbf{x}) & \\ & & h_3^{-2}(\mathbf{x}) \end{bmatrix}.$$

$\mathcal{R}(\mathbf{x})$ is an orthonormal matrix providing the local orientation given by the eigenvectors $(\mathbf{v}_i(\mathbf{x}))_{i=1,3}$ and $(\lambda_i(\mathbf{x}))_{i=1,3}$ are the local eigenvalues.

DEFINITION 3.9 (Directional local mesh sizes).– $(h_i(\mathbf{x}))_{i=1,3} = (\lambda_i(\mathbf{x})^{-\frac{1}{2}})_{i=1,3}$ *are the local mesh sizes along the principal directions of* $\mathbf{M}$.

DEFINITION 3.10 (Local unit ball).– *The local unit ball of* $\mathcal{M}$ *at point* $\mathbf{x}$ *is defined by*

$$\mathcal{B}_{\mathcal{M}(\mathbf{x})} = \mathbf{x} + \Phi_{\mathcal{M}(\mathbf{x})}(1), \qquad [3.13]$$

where $\Phi_{\mathcal{M}}(1)$ *is defined by [3.4] (see also Figure 3.5).*

DEFINITION 3.11 (Density of a metric).– *The density* d *of* $\mathbf{M}$ *is defined from its eigenvalues as*

$$d(\mathbf{x}) = det\big(\mathcal{M}(\mathbf{x})\big)^{\frac{1}{2}} = (\lambda_1(\mathbf{x})\,\lambda_2(\mathbf{x})\,\lambda_3(\mathbf{x}))^{\frac{1}{2}} = (h_1(\mathbf{x})\,h_2(\mathbf{x})\,h_3(\mathbf{x}))^{-1}.$$

DEFINITION 3.12 (Anisotropic quotients of a metric).– *The anisotropic quotients* r_i *of metric* $\mathbf{M}$ *are defined as* $r_i(\mathbf{x}) = h_i(\mathbf{x})^3\,(h_1(\mathbf{x})\,h_2(\mathbf{x})\,h_3(\mathbf{x}))^{-1}$.

DEFINITION 3.13 (Anisotropic ratios of a metric).– *The anisotropic ratios* $\bar{r}_i$ *of metric* $\mathbf{M}$ *are defined as* $\bar{r}_i(\mathbf{x}) = \dfrac{max_i h_i(\mathbf{x})}{min_i h_i(\mathbf{x})}$.

We can easily observe the following:

PROPOSITION 3.2.– $\mathbf{M} = (\mathcal{M}(\mathbf{x}))_{\mathbf{x}\in\Omega}$ locally writes

$$\mathcal{M}(\mathbf{x}) = d^{\frac{2}{3}}(\mathbf{x})\,\mathcal{R}(\mathbf{x}) \begin{bmatrix} r_1^{-\frac{2}{3}}(\mathbf{x}) & & \\ & r_2^{-\frac{2}{3}}(\mathbf{x}) & \\ & & r_3^{-\frac{2}{3}}(\mathbf{x}) \end{bmatrix} {}^t\mathcal{R}(\mathbf{x}). \square$$

The density d controls only the local level of accuracy of $\mathbf{M}$. Increasing or decreasing d does not change the anisotropic properties or the orientation (see Figure 3.12, left). Anisotropic quotients arise from the quotient of different parallelepipeds (see Figure, 3.12 right). We also define the *complexity* $\mathcal{C}$ of $\mathbf{M}$:

$$\mathcal{C}(\mathbf{M}) = \int_\Omega d(\mathbf{x})\,dx = \int_\Omega \sqrt{\det(\mathcal{M}(\mathbf{x}))}\,dx.$$

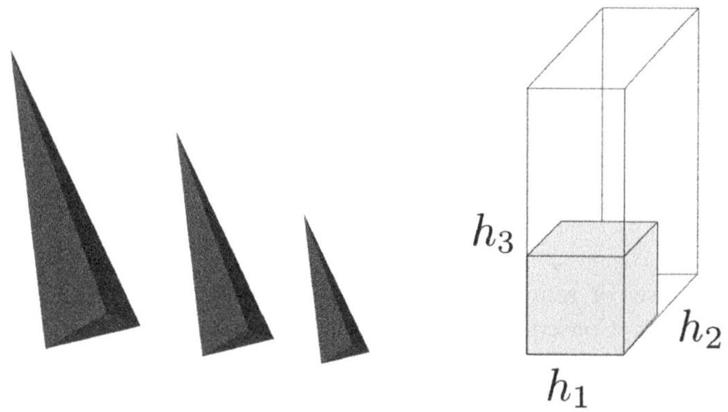

Figure 3.12. *Left: Different unit elements where only the density increases from left to right. Right: The geometric interpretation of anisotropic quotients as quotients of parallelepipeds volumes. For a color version of this figure, see www.iste.co.uk/dervieux/meshadaptation1*

This real-value parameter is useful to quantify the global level of accuracy of $(\mathcal{M}(\mathbf{x}))_{\mathbf{x}\in\Omega}$. It can also be interpreted as the continuous counterpart of the *number of vertices of a discrete mesh* while d can be interpreted as the continuous counterpart of the *number of vertices per volume unit of a discrete mesh*. This quantity also leads to the definition of a sequence of embedded Riemannian spaces. Two embedded Riemannian spaces have the same anisotropic ratios and orientations. They differ only in their complexity.

DEFINITION 3.14 (Embedded Riemannian spaces).– *Two Riemannian spaces, say,* $(\mathcal{M}(\mathbf{x}))_{\mathbf{x}\in\Omega}$ *and* $(\mathcal{N}(\mathbf{x}))_{\mathbf{x}\in\Omega}$*, are embedded if a constant c exists such that*

$$\forall \mathbf{x} \in \Omega, \quad \mathcal{N}(\mathbf{x}) = c\,\mathcal{M}(\mathbf{x}).$$

Conversely, from $\mathbf{M} = (\mathcal{M}(\mathbf{x}))_{\mathbf{x}\in\Omega}$, we can deduce $\mathbf{N} = (\mathcal{N}(\mathbf{x}))_{\mathbf{x}\in\Omega}$ of prescribed complexity N having the same anisotropic properties (anisotropic orientations and ratios) by considering

$$\mathcal{N}(\mathbf{x}) = \left(\frac{N}{\mathcal{C}(\mathbf{M})}\right)^{\frac{2}{3}} \mathcal{M}(\mathbf{x}).$$

In the context of error estimation, this notion enables convergence order study with respect to an increasing complexity N. Consequently, the complexity $\mathcal{C}(\mathbf{M})$ is also the continuous counterpart of the classical parameter h used for uniform meshes while studying convergence. In the continuous framework, the uniform refinement consisting of dividing by two each edge of a uniform mesh of size h is given as

$$\mathcal{M}_i = 4^i \begin{bmatrix} \frac{1}{h^2} & & \\ & \frac{1}{h^2} & \\ & & \frac{1}{h^2} \end{bmatrix},$$

where i is the level of refinement. $(\mathcal{M}_i)_{i=1...k}$ defines a sequence of embedded Riemannian spaces. Consequently, this simple practical adaptive strategy has a simple continuous interpretation in terms of embedded Riemannian spaces. However, when dealing with anisotropic meshes, a unique size h is no more sufficient to give quantitive information on the accuracy. The size h is then replaced by the complexity.

3.5.3.1. *Unit mesh*

Defining a unit mesh is more delicate than defining a unit element because the existence of a mesh composed only of unit regular simplexes with respect to a given Riemannian metric space is not guaranteed. For instance, if the Riemannian metric space is not compatible with the domain size, then it clearly does not exist such a discrete mesh. To avoid this problem, let us look at the existence of a discrete mesh only made of unit regular simplexes with respect to a Riemannian metric space in $\mathbb{R}^n$. To further simplify the problem, we first consider the Euclidean space $(\mathcal{I}_n(\mathbf{x}))_{\mathbf{x}\in\mathbb{R}^n}$.

It is well known that $\mathbb{R}^3$ cannot be filled only with the regular tetrahedron while it is possible to fill $\mathbb{R}^2$ with the equilateral triangle. Consequently, even for the simplest continuous mesh $(\mathcal{I}_3(\mathbf{x}))_{\mathbf{x}\in\mathbb{R}^3}$, there is no discrete mesh composed only of the unit regular tetrahedron. Therefore, the notion of unit mesh has to be relaxed.

DEFINITION 3.15 (Unit mesh).– *A discrete mesh $\mathcal{H}$ of a domain $\Omega \subset \mathbb{R}^n$ is unit for Riemannian metric space $(\mathcal{M}(\mathbf{x}))_{\mathbf{x}\in\Omega}$ if all its elements are quasi-unit.*

The analogs for the unit mesh of the local unit ball and the local mesh size of the metric are written as follows:

DEFINITION 3.16 (Local unit ball in a mesh).– *Let i be a vertex i of coordinates $\mathbf{x}_i$ of the mesh $\mathcal{H}$. The local unit ball $\mathcal{B}_{\mathcal{H}}$ of i is the set of elements having i as vertex.*

Now, let us give a meaning to quasi-unit in three dimensions. A first way to relax the definition of unity is to take into account technical constraints imposed by mesh generators. To converge (and to avoid cycling) while analyzing edges length, the meshing algorithm considers an admissible edge length interval of the form $[\frac{1}{\alpha}, \alpha]$

with $\alpha > 0$ (see (Frey 2001)). If the symmetry property is required, that is, $\frac{\alpha}{2} = \frac{1}{\alpha}$, then we obtain $\alpha = \sqrt{2}$. Therefore, with regard to the meshing requirement, a tetrahedron K defined by its list of edges $(\mathbf{e}_i)_{i=1\ldots6}$ is said to be quasi-unit if $\forall i \in [1,6]$, $\ell_{\mathcal{M}}(\mathbf{e}_i) \in [\frac{1}{\sqrt{2}}, \sqrt{2}]$. Nevertheless, we do not know if this definition provides the existence of a unit mesh for the Euclidean space $(\mathcal{I}_3(\mathbf{x}))_{\mathbf{x}\in\mathbb{R}^3}$. In the following, this question of existence is studied by means of the space filling tetrahedra.

Non-regular space filling tetrahedra. The study of space-filling tetrahedra is an old geometrical question (Naylor 1999; Senechal 1981). In the past, it has been demonstrated that there exist sets of non-regular space-filling tetrahedra: the Sommerville tetrahedra (Sommerville 1923) and the Goldberg tetrahedra families (Goldberg 1974).

The Sommerville tetrahedra are built on particular splittings of the unit cube (see Figure 3.13). We recall these tetrahedra because of their vertices coordinates. Actually, only the last vertex distinguishes them. K is denoted $(\mathbf{v_1}, \mathbf{v_2}, \mathbf{v_3}, \mathbf{v_4})$ with $\mathbf{v_1} = (0,0,0)$, $\mathbf{v_2} = (\frac{1}{2}, -\frac{1}{2}, \frac{1}{2})$, $\mathbf{v_3} = (\frac{1}{2}, \frac{1}{2}, \frac{1}{2})$ and

- $\mathbf{v_4} = (\frac{1}{2}, 0, 0)$ for the Sommerville tetrahedron 1;
- $\mathbf{v_4} = (1, 0, 0)$ for the Sommerville tetrahedron 2;
- $\mathbf{v_4} = (\frac{1}{2}, -\frac{1}{2}, -\frac{1}{2})$ for the Sommerville tetrahedron 3;
- $\mathbf{v_4} = (\frac{1}{2}, 0, -\frac{1}{4})$ for the Sommerville tetrahedron 4.

The Goldberg tetrahedra are based on the splitting of a prism, the basis of which is the equilateral triangle (see Figure 3.14). Their coordinates depend on an initial choice of two lengths a and e. We specify one of the Goldberg tetrahedra for the specific choice $a = \frac{1}{3}$ and $e = 1$:

– $\mathbf{v_1} = (0,0,0)$, $\mathbf{v_2} = (0,0,1)$, $\mathbf{v_3} = (0,1,\frac{1}{3})$ and $\mathbf{v_4} = (\frac{\sqrt{3}}{2}, \frac{1}{2}, \frac{2}{3})$ for the Goldberg tetrahedron.

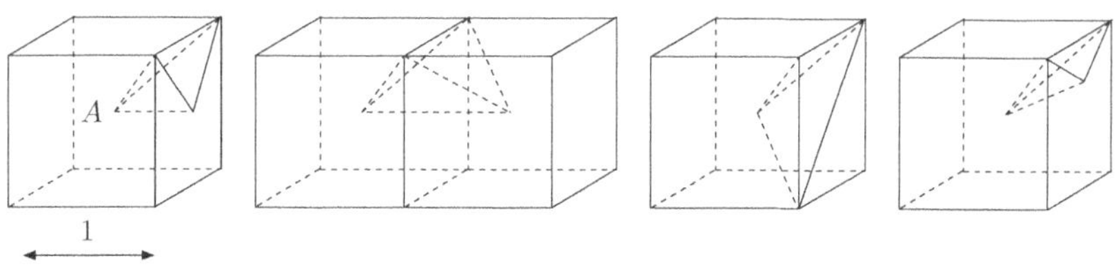

Figure 3.13. *From left to right, Sommerville tetrahedra* 1, 2, 3 *and* 4

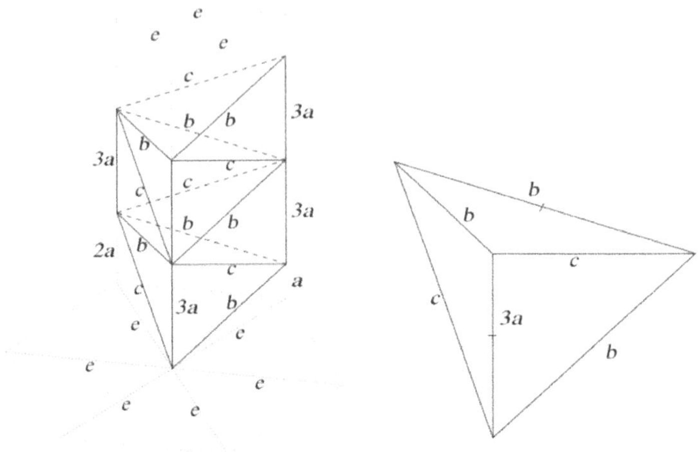

Figure 3.14. *Goldberg's tetrahedra family. They are parameterized by the prescription of the lengths a and e, with $b^2 = a^2 + e^2$ and $c^2 = 4a^2 + e^2$. Left: gathering tetrahedra together fills a prism, the basis of which is the regular triangle of side length e*

We suggest now comparing these space-filling tetrahedra to the unit regular tetrahedron. To this end, all these tetrahedra are scaled such that their volumes are equal to $\frac{\sqrt{2}}{12}$. The resulting edge lengths for each tetrahedron are specified in Table 3.2. We notice that the proposed notion of quasi-unit element is verified for the Sommerville tetrahedra 1 and 2, and the Goldberg tetrahedron. Therefore, there exists space filling tetrahedra that are quasi-units for the metric $\mathcal{I}_3$ in the sense proposed above.

Controlling the volume. Unfortunately, the weaker constraint on the edges length can lead to the generation of quasi-unit elements with a null volume. For instance, in $(\mathbb{R}^3, \mathcal{I}_3)$, the regular tetrahedron with edges length equal to $\sqrt{2}$ is quasi-unit for $\mathcal{I}_3$. However, if one vertex is projected orthogonally on the opposite face, then a quasi-unit element of null volume is obtained. Indeed, three edges are of length $\sqrt{2}$ and the three other are of length $\frac{\sqrt{6}}{3} \approx 0.816 \in \left[\frac{1}{\sqrt{2}}, \sqrt{2}\right]$. In consequence, controlling only the edges length is not sufficient, the volume must also be controlled to relax the notion of unit element. Practically, this is achieved by managing a quality function, defined as follows:

DEFINITION 3.17.– *The quality function of an element K is given by the following quantity:*

$$Q_{\mathcal{M}}(K) = \frac{36}{3^{\frac{1}{3}}} \frac{|K|_{\mathcal{M}}^{\frac{2}{3}}}{\sum_{i=1}^{6} \ell_{\mathcal{M}}^2(\mathbf{e_i})} \in [0, 1]. \qquad [3.14]$$

REMARK 3.1.– Instead of $Q_{\mathcal{M}}$, the quality function $Q_{\mathcal{M}}^{-1} = \frac{1}{Q_{\mathcal{M}}}$ can be considered. As the variation range becomes $[1, \infty[$, the discrimination of bad elements is made easier. □

For the perfect regular tetrahedron, whatever its edges length, the quality function is equal to 1. For a null volume tetrahedron, $Q_{\mathcal{M}}$ is 0. The qualities of the space filling tetrahedra are given in Table 3.2. Note that $Q_{\mathcal{M}}$ only quantifies the gap to the regular tetrahedron shape. We deduce the following definition of quasi-unit element, which is also practically used by mesh generators,

DEFINITION 3.18.– *A tetrahedron K defined by its list of edges $(\mathbf{e_i})_{i=1...6}$ is said quasi-unit for $\mathcal{M}$ if*

$$\forall i \in [1, 6], \quad \ell_{\mathcal{M}}(\mathbf{e_i}) \in \left[\frac{1}{\sqrt{2}}, \sqrt{2}\right] \quad \textit{and} \quad Q_{\mathcal{M}}(K) \in [\alpha, 1] \ \textit{with} \ \alpha > 0.$$

In our case, $\alpha = 0.8$ is an acceptable value as it enables the Sommerville tetrahedra 1 and 2, and the Goldberg tetrahedron to be generated. In practice, we use $\alpha = 2$.

3.5.3.2. *Generating an adapted mesh*

To summarize the previous analysis, if $(\mathcal{M}(\mathbf{x}))_{\mathbf{x}\in\Omega}$ is the prescribed Riemannian metric space, then the adaptive mesher (tries to) generates a unit mesh of domain Ω with respect to $(\mathcal{M}(\mathbf{x}))_{\mathbf{x}\in\Omega}$ according to definitions 3.15 and 3.18. The quality of the generated adapted mesh with respect to the perfect unit mesh can be analyzed in terms of size and shape by computing, respectively, the length of the edges and the quality of the elements in the metric. Consequently, the generated mesh is uniform and unit in the Riemannian metric space, while it is adapted and anisotropic in the Euclidean space. The following pedagogic example gives an intuition of this process.

EXAMPLE 3.3.– Figure 3.15 shows in background the France map in the meter metric and in foreground the France map in the train time metric, that is, the time spent by the train to go from one place to another. It shows how a uniform Cartesian grid on the France map in the meter metric is deformed after being mapped onto the France map in the time metric. Exactly the same idea is used in the generation of adapted anisotropic meshes.

Tetrahedron	Coeff.	Edges length						Quality $Q_{\mathcal{M}}(K)$	$Q^{-1}_{\mathcal{M}}(K)$
Sommerville 1	$\sqrt{2}$	0.70	1.22	1.22	1.0	1.0	1.41	0.800	1.25
Sommerville 2	$2^{\frac{1}{6}}$	1.12	0.970	0.970	0.970	0.970	1.12	**0.954**	**1.04821**...
Sommerville 3	$2^{\frac{1}{6}}$	0.970	0.970	0.970	1.12	**1.59**	1.12	0.763	1.31061...
Sommerville 4	$12^{-\frac{1}{3}}2^{\frac{3}{2}}$	**0.691**	1.07	1.07	1.12	1.12	1.23	0.886	1.12866...
Goldberg	$3^{-\frac{1}{2}}2^{\frac{1}{6}}$	0.932	0.990	1.12	1.12	0.990	0.990	0.950	1.05263...

Table 3.2. *Space filling tetrahedra characteristics. Coeff. is the coefficient that scales the tetrahedron onto a unit volume tetrahedron K_1, that is, $|K_1| = \sqrt{2}/12$. The edges length and the tetrahedron quality $Q_{\mathcal{M}}(K)$ given by formula [3.14] are provided (largest values are in bold)*

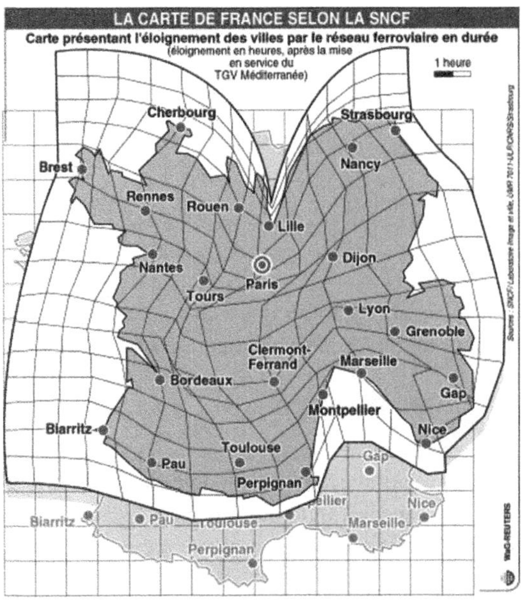

Figure 3.15. *A picture (from SNCF) representing the France map in the time metric, that is, the time required to go from one place to another. For a color version of this figure, see www.iste.co.uk/dervieux/meshadaptation1*

Let us say a few words about the size prescription by a given Riemannian metric field. We have seen that the size of the unit ball associated with metric tensor $\mathcal{M}$ in the

ith eigenvector direction is given by $h_i = \lambda_i^{-\frac{1}{2}}$. We then deduce that the prescribed size in direction $\mathbf{e}$ is

$$h_{\mathcal{M}}(\mathbf{e}) = \frac{\|\mathbf{e}\|_2}{\ell_{\mathcal{M}}(\mathbf{e})}.$$

In the case of an Euclidean metric space, the previous relation simply reads

$$h_{\mathcal{M}}(\mathbf{e}) = \frac{\|\mathbf{e}\|_2}{\sqrt{{}^t\mathbf{e}\mathcal{M}\mathbf{e}}}. \qquad [3.15]$$

3.5.4. *Quantifying mesh anisotropy*

In three dimensions, mesh anisotropy can be quantified by two notions:

– the *anisotropic ratios*;

– the *anisotropic quotients*.

We first recall both notions and the way they are evaluated numerically. Deriving these quantities for an element relies on the fact that there always exists a unique metric tensor for which this element is unit. If $\mathcal{M}_K$ denotes the metric tensor associated with element K, solving the following linear system provides $\mathcal{M}_K$:

$$(S)\ \begin{cases} \ell^2_{\mathcal{M}_K}(\mathbf{e}_1) = 1 \\ \vdots \\ \ell^2_{\mathcal{M}_K}(\mathbf{e}_6) = 1, \end{cases}$$

where $(\mathbf{e}_i)_{i=1,6}$ is the edges list of K and $\ell^2_{\mathcal{M}_K}(\mathbf{e}_i) = {}^t\mathbf{e}_i\,\mathcal{M}_K\,\mathbf{e}_i$. (S) admits a unique solution as soon as the volume of K is not null.

DEFINITION 3.19 (Anisotropic ratio/quotient of an element).– *Once $\mathcal{M}_K$ associated with element K is computed, the anisotropic ratio and the anisotropic quotient associated with element K are simply given by*

$$\textit{ratio} = \sqrt{\frac{\min_i \lambda_i}{\max_i \lambda_i}} = \frac{\max_i h_i}{\min_i h_i}, \quad \textit{and} \quad \textit{quotient} = \frac{\max_i h_i^3}{h_1 h_2 h_3},$$

where $(\lambda_i)_{i=1,3}$ are the eigenvalues of $\mathcal{M}_K$ and $(h_i)_{i=1,3}$ are the corresponding sizes.

The anisotropic ratio stands for the maximum elongation of a tetrahedron by comparing two eigen directions. The anisotropic quotient represents the overall anisotropic ratio of a tetrahedron taking into account all the possible directions. It corresponds to the overall gain in three dimensions of *an anisotropic adapted mesh* as compared to *an isotropic adapted one*. The gain is of course even greater when compared to a uniform mesh.

Back to the introducing example. We revisit here the introduction example of finding an optimal mesh to represent the half-cylinder function defined on $[-1,1] \times [-1,1]$. We recall that the solution to get an optimal mesh consists in putting all the vertices along $y = -1$ and $y = 1$. Consequently, we impose a mesh size dx along the x-axis and a size of 1 along the y-axis. dx is chosen so that we reach the desired number of vertices. So $dx = 2/71$ to get 144 vertices. By a direct applications of the definitions of Euclidean metric-space, the metric prescribing theses sizes is given by

$$\mathcal{M} = \begin{pmatrix} \frac{1}{dx^2} & 0 \\ 0 & \frac{1}{2^2} \end{pmatrix}.$$

We verify that if a mesh generator works in $([-1,1]\times[-1,1], \mathcal{M})$, then the optimal mesh built by hand is unit with respect to this metric. First, let us compute the size of an edge aligned along the x-axis: we have $\|{}^t(x,0)\|^2_{\mathcal{M}} = x^2/dx^2$; this length is unit when $x = dx$. Along the y-axis, we have $\|{}^t(0,y)\|_{\mathcal{M}} = y^2/2^2$; this length is unit when $y = 2$. The optimal mesh is then unit with respect to $\mathcal{M}$, we can also verify that the area of triangle in the metric is $\sqrt{3}/4$.

3.6. Operations on metrics

The main advantage when working with metric spaces is the well-posedness of operations on metric tensors, among which are the metric intersection and the metric interpolation. These operations have a straightforward geometric interpretation when considering the ellipsoid associated with a metric.

3.6.1. *Metric intersection*

When several metrics are specified at a point of the domain, all these metric tensors must be reduced to a single one due to mesh generation concerns. The *metric intersection* consists of choosing in any direction the most restrictive size constraint imposed by this set of metrics.

Formally speaking, let $\mathcal{M}_1$ and $\mathcal{M}_2$ be two metric tensors given at a point. The metric tensor $\mathcal{M}_{1\cap 2}$ corresponding to the intersection of $\mathcal{M}_1$ and $\mathcal{M}_2$ is the one prescribing the largest possible size under the constraint that the size in each direction is always smaller than the sizes prescribed by $\mathcal{M}_1$ and $\mathcal{M}_2$. Let us give a geometric interpretation of this operator. Metric tensors are geometrically represented by an ellipse in 2D and an ellipsoid in 3D. But the intersection between two metrics is not directly the intersection between two ellipsoids as their geometric intersection is not an ellipsoid. Therefore, we seek for the largest ellipsoid representing $\mathcal{M}_{1\cap 2}$ included in the geometric intersection of the ellipsoids associated with $\mathcal{M}_1$ and $\mathcal{M}_2$ (see Figure 3.16, left). The ellipsoid (metric) verifying this property is obtained by using the simultaneous reduction of two metrics.

Simultaneous reduction. Simultaneous reduction enables us to find a common basis $(\mathbf{e}_1, \mathbf{e}_2, \mathbf{e}_3)$ such that $\mathcal{M}_1$ and $\mathcal{M}_2$ are congruent to a diagonal matrix in this basis, and then to deduce the intersected metric. To do so, the matrix $\mathcal{N} = \mathcal{M}_1^{-1}\mathcal{M}_2$ is introduced. $\mathcal{N}$ is diagonalizable with real-eigenvalues. The normalized eigenvectors of $\mathcal{N}$ denoted by $\mathbf{e}_1$, $\mathbf{e}_2$ and $\mathbf{e}_3$ constitute a common diagonalization basis for $\mathcal{M}_1$ and $\mathcal{M}_2$. The entries of the diagonal matrices, that are associated with the metrics $\mathcal{M}_1$ and $\mathcal{M}_2$ in this basis, are obtained with the Rayleigh formula[2]:

$$\lambda_i = {}^t\mathbf{e}_i \mathcal{M}_1 \mathbf{e}_i \quad \text{and} \quad \mu_i = {}^t\mathbf{e}_i \mathcal{M}_2 \mathbf{e}_i, \quad \text{for} \quad i = 1 \ldots 3.$$

Let $\mathcal{P} = (\mathbf{e}_1\, \mathbf{e}_2\, \mathbf{e}_3)$ be the matrix the columns of which are the eigenvectors $\{\mathbf{e}_i\}_{i=1\ldots3}$ of $\mathcal{N}$. $\mathcal{P}$ is invertible as $(\mathbf{e}_1, \mathbf{e}_2, \mathbf{e}_3)$ is a basis of $\mathbb{R}^3$. We have

$$\mathcal{M}_1 = {}^t\mathcal{P}^{-1} \begin{pmatrix} \lambda_1 & 0 & 0 \\ 0 & \lambda_2 & 0 \\ 0 & 0 & \lambda_3 \end{pmatrix} \mathcal{P}^{-1} \quad \text{and} \quad \mathcal{M}_2 = {}^t\mathcal{P}^{-1} \begin{pmatrix} \mu_1 & 0 & 0 \\ 0 & \mu_2 & 0 \\ 0 & 0 & \mu_3 \end{pmatrix} \mathcal{P}^{-1}.$$

Computing the metric intersection. The resulting intersected metric $\mathcal{M}_{1\cap 2}$ is then analytically given by

$$\mathcal{M}_{1\cap 2} = \mathcal{M}_1 \cap \mathcal{M}_2 = {}^t\mathcal{P}^{-1} \begin{pmatrix} \max(\lambda_1, \mu_1) & 0 & 0 \\ 0 & \max(\lambda_2, \mu_2) & 0 \\ 0 & 0 & \max(\lambda_3, \mu_3) \end{pmatrix} \mathcal{P}^{-1}.$$

The ellipsoid associated with $\mathcal{M}_{1\cap 2}$ is the largest ellipsoid included in the geometric intersection region of the ellipsoids associated with $\mathcal{M}_1$ and $\mathcal{M}_2$, and the proof is given in Alauzet and Frey (2003). Numerically, to compute $\mathcal{M}_{1\cap 2}$, the real-eigenvalues of $\mathcal{N}$ are first evaluated with a Newton algorithm. Then, the eigenvectors of $\mathcal{N}$, which define $\mathcal{P}$, are computed using the fact that an eigenvector lies in the orthogonal of $Im(\mathcal{M} - \lambda Id)$ and that the three vectors are orthonormal.

2 λ_i and μ_i are not the eigenvalues of $\mathcal{M}_1$ and $\mathcal{M}_2$. They are spectral values associated with basis $(\mathbf{e}_1, \mathbf{e}_2, \mathbf{e}_3)$.

REMARK 3.2.– The intersection operation is not commutative. Consequently, when more than two metrics are intersected, the result depends on the order of intersection. In this case, the resulting intersected metric is not anymore optimal. If we seek for the largest ellipsoid included in the geometric intersection region of several (> 2) metrics, then the John ellipsoid has to be found by solving an optimization problem (Loseille 2008).

3.6.2. *Metric interpolation*

In practice, the metric field is only known discretely at mesh vertices. The definition of an interpolation procedure on metrics is therefore mandatory to be able to compute the metric at any point of the domain. For instance, the computation of the volume of an element using quadrature formula with relation [3.6] requires the computation of some interpolated metrics inside the considered element.

Several interpolation schemes have been proposed in Alauzet and Frey (2003), which are based on the simultaneous reduction. The main drawback of these approaches is that the interpolation operation is not commutative. Hence, the result depends on the order in which the operations are performed when more than two metrics are involved. Moreover, such interpolation schemes do not satisfy useful properties such as the maximum principle. Consequently, to design an interpolation scheme on these objects, one needs a consistent operational framework. We suggest to consider the log-Euclidean framework introduced in Arsigny et al. (2006).

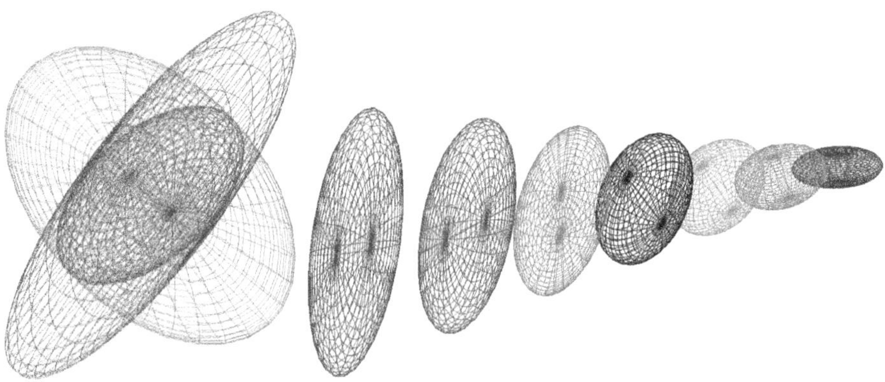

Figure 3.16. *Left view illustrating the metric intersection procedure with the simultaneous reduction in three dimensions. In red, the resulting metric of the intersection of the blue and green metrics. Right, metric interpolation along a segment where the endpoints metrics are the blue and violet ones. For a color version of this figure, see www.iste.co.uk/dervieux/meshadaptation1*

Log-Euclidean framework. We first define the notion of metric logarithm and matrix exponential. The *metric logarithm* is defined on the set of metric tensors. For metric tensor $\mathcal{M} = \mathcal{R}\Lambda^t\mathcal{R}$, it is given by

$$\ln(\mathcal{M}) := \mathcal{R}\ln(\Lambda)^t\mathcal{R},$$

where $\ln(\Lambda) = diag(\ln(\lambda_i))$. The *matrix exponential* is defined on the set of symmetric matrices. For any symmetric matrix $\mathcal{S} = \mathcal{Q}\Xi^t\mathcal{Q}$, it is given by

$$\exp(\mathcal{S}) := \mathcal{Q}\exp(\Xi)^t\mathcal{Q},$$

where $\exp(\Xi) = diag(\exp(\xi_i))$. We can now define the *logarithmic addition* $\oplus$ and the *logarithmic scalar multiplication* $\odot$:

$$\begin{aligned} \mathcal{M}_1 \oplus \mathcal{M}_2 &:= \exp\left(\ln(\mathcal{M}_1) + \ln(\mathcal{M}_2)\right) \\ \alpha \odot \mathcal{M} &:= \exp\left(\alpha.\ln(\mathcal{M})\right) = \mathcal{M}^\alpha. \end{aligned}$$

The logarithmic addition is commutative and coincides with matrix multiplication whenever the two tensors $\mathcal{M}_1$ and $\mathcal{M}_2$ commute in the matrix sense. The space of metric tensors, supplied with the logarithmic addition $\oplus$ and the logarithmic scalar multiplication $\odot$ is a vector space.

REMARK 3.3.– This framework allows more general computations to be carried out on metric tensors, such as statistical studying or the resolution of PDEs on metric tensors.

Metric interpolation in the log-Euclidean framework. We propose to use the linear interpolation operator derived from the log-Euclidean framework. Let $(\mathbf{x}_i)_{i=1...k}$ be a set of vertices and $(\mathcal{M}(\mathbf{x}_i))_{i=1...k}$ their associated metrics. Then, for a point $\mathbf{x}$ of the domain such that

$$\mathbf{x} = \sum_{i=1}^{k} \alpha_i.\mathbf{x}_i \quad \text{with} \quad \sum_{i=1}^{k} \alpha_i = 1,$$

the interpolated metric is defined by

$$\mathcal{M}(\mathbf{x}) = \exp\left(\sum_{i=1}^{k} \alpha_i \ln(\mathcal{M}(\mathbf{x}_i))\right). \qquad [3.16]$$

This interpolation is commutative, but its bottleneck is to perform k diagonalizations and to request the use of the logarithm and the exponential functions which are CPU consuming. However, this procedure is essential to define continuously the metric map on the entire domain. Moreover, it has been demonstrated in Arsigny et al. (2006) that this interpolation preserves the maximum principle, that is, for an edge **pq** with endpoints metrics $\mathcal{M}(\mathbf{p})$ and $\mathcal{M}(\mathbf{q})$ such that $\det(\mathcal{M}(\mathbf{p})) < \det(\mathcal{M}(\mathbf{q}))$, then we have $\det(\mathcal{M}(\mathbf{p})) < \det(\mathcal{M}(\mathbf{p} + t\,\mathbf{pq})) < \det(\mathcal{M}(\mathbf{q}))$ for all $t \in [0, 1]$. Accordingly, this metric interpolation enables a continuous metric field to be defined throughout the entire discretized domain. When a metric is required at a point, we determine to which element the point belongs. Then, we apply relation [3.16], where α_i are the barycentric coordinates of the point with respect to the element. Figure 3.16 illustrates metric interpolation along a segment, for which the initial data are the endpoints metrics.

REMARK 3.4.– The interpolation formulation [3.16] reduces to

$$\mathcal{M}(\mathbf{x}) = \prod_{i=1}^{k} \mathcal{M}(\mathbf{x}_i)^{\alpha_i},$$

if all the metrics commute. Therefore, an arithmetic mean in the log-Euclidean framework could be interpreted as a geometric mean in the space of metric tensors.

3.7. Computation of geometric quantities

Let $\mathcal{M}$ be a discrete metric field defined at the vertices of a mesh $\mathcal{H}$ of a domain Ω_h. Because of the interpolation operation, we have a continuous metric field in the whole domain, that is, a Riemannian metric space $(\mathcal{M}(\mathbf{x}))_{\mathbf{x}\in\Omega_h}$. This representation of the metric field depends on $\mathcal{H}$ as the interpolation law is applied at the element level.

3.7.1. *Computation of lengths*

Let $\mathbf{a}$ and $\mathbf{b}$ be two points of the domain. The length of segment $\mathbf{ab}$ is given by integral [3.5]. This integral can be numerically approximated using a k-points Gaussian quadrature with weights $(\omega_i)_{i=1...k}$ and barycentric coefficients $(\alpha_i)_{i=1...k}$:

$$\ell_{\mathcal{M}}(\mathbf{ab}) = \int_0^1 \sqrt{{}^t\mathbf{ab}\,\mathcal{M}(\mathbf{a} + t\,\mathbf{ab})\,\mathbf{ab}}\,dt \approx \sum_{i=1}^{k} \omega_i \sqrt{{}^t\mathbf{ab}\,\mathcal{M}(\mathbf{a} + \alpha_i\,\mathbf{ab})\,\mathbf{ab}},$$

where $\mathcal{M}(\mathbf{a} + \alpha_i\,\mathbf{ab})$ is the metric at the ith Gauss point.

In the context of meshing, we are interested in the evaluation of edge length in the metric. This is a recurrent operation, which could be time consuming for the mesh generator. Fortunately, this operation is at the edge level where the metric is known at the endpoints. The problem can thus be simplified by considering a variation law of the metric along the edge and computing the edge length in the metric analytically. Formally speaking, let $\mathbf{e} = \mathbf{p_1 p_2}$ be an edge of the mesh of Euclidean length $\|\mathbf{e}\|_2$, and $\mathcal{M}(\mathbf{p}_1)$ and $\mathcal{M}(\mathbf{p}_2)$ be the metrics at the edge extremities $\mathbf{p}_1$ and $\mathbf{p}_2$. We denote by $\ell_i(\mathbf{e}) = \sqrt{{}^t\mathbf{e}\mathcal{M}(\mathbf{p}_i)\mathbf{e}}$ the length of the edge in metric $\mathcal{M}(\mathbf{p}_i)$. We assume $\ell_1(\mathbf{e}) > \ell_2(\mathbf{e})$ and we set $a = \frac{\ell_1(\mathbf{e})}{\ell_2(\mathbf{e})}$. Now, we consider the metric variation law associated with the interpolation operator on metrics given in section 3.6.2. The restriction of the (multi-dimensional) metric interpolation operator given by relation [3.16] to an edge $\mathbf{e} = \mathbf{p_1 p_2}$ leads to

$$\lambda(t) = \exp\left((1-t)\ln(\lambda_1) + t\ln(\lambda_2)\right) = \lambda_1^{(1-t)}\lambda_2^{t}$$

$$\Longleftrightarrow \frac{1}{h^2(t)} = \exp\left(\ln\left(\frac{1}{h_1^{2(1-t)}\,h_2^{2t}}\right)\right) \Longleftrightarrow h(t) = h_1^{(1-t)}\,h_2^{t}.$$

Consequently, this multi-dimensional operator is equivalent to a geometric interpolation law on h on the edge. Note that the geometric interpolation on λ is equivalent to the geometric interpolation on h.

PROPOSITION 3.3.– Assuming that metric field follows a geometric variation law, that is, a geometric interpolation on h: $h(t) = h_1^{(1-t)}h_2^{t}$, then we have

$$\ell_{\mathcal{M}}(\mathbf{e}) = \ell_1(\mathbf{e})\frac{a-1}{a\ln(a)} \quad \text{with } a = \ell_1(\mathbf{e})/\ell_2(\mathbf{e}).$$

PROOF.– According to relation [3.15], we have

$$\ell_1(\mathbf{e}) = \frac{\|\mathbf{e}\|_2}{h_1(\mathbf{e})}, \quad \ell_2(\mathbf{e}) = \frac{\|\mathbf{e}\|_2}{h_2(\mathbf{e})}, \quad a = \frac{h_2(\mathbf{e})}{h_1(\mathbf{e})},$$

and

$$\ell_{\mathcal{M}}(\mathbf{e}) = \int_0^1 \sqrt{{}^t\mathbf{e}\,\mathcal{M}(t)\,\mathbf{e}}\,\mathrm{d}t = \|\mathbf{e}\|_2 \int_0^1 \frac{1}{h_{\mathcal{M}(t)}(\mathbf{e})}\mathrm{d}t\,,$$

where we have denoted in short $\mathcal{M}(t) = \mathcal{M}(\mathbf{p}_1 + t\, \mathbf{p}_1\mathbf{p}_2)$. For the geometric law on h: $h_{\mathcal{M}(t)} = h_1^{(1-t)} h_2^t$, it yields

$$\ell_{\mathcal{M}}(\mathbf{e}) = \|\mathbf{e}\|_2 \int_0^1 \frac{1}{h_1^{(1-t)}(\mathbf{e})\, h_2^t(\mathbf{e})}\, \mathrm{d}t = \|\mathbf{e}\|_2 \int_0^1 \frac{1}{h_1(\mathbf{e}) a^t}\, \mathrm{d}t = \ell_1(\mathbf{e}) \int_0^1 a^{-t}\, \mathrm{d}t$$

$$= \ell_1(\mathbf{e}) \left[-\frac{a^{-t}}{\ln(a)} \right]_0^1 = \ell_1(\mathbf{e}) \frac{a-1}{a \ln(a)}. \square$$

3.7.2. *Computation of volumes*

The evaluation of a tetrahedron volume in a Riemannian metric space consists of computing numerically integral [3.6]. If a first-order approximation in the log-Euclidean framework is considered, we get

$$|K|_{\mathcal{M}} = \int_K \sqrt{\det \mathcal{M}} \mathrm{d}\mathbf{x} \approx \sqrt{\det\left(\exp\left(\frac{1}{4} \sum_{i=1}^{4} \log\left(\mathcal{M}_i\right) \right) \right)}\ |K|_{\mathcal{I}_3}.$$

Higher-order approximation can be obtained by using Gaussian quadrature. For instance, if one considers a k-point Gaussian quadrature with weights $(\omega_j)_{j=1\ldots k}$ and barycentric coefficients $(\beta_j^1, \beta_j^2, \beta_j^3, \beta_j^4)_{j=1\ldots k}$, it yields

$$|K|_{\mathcal{M}} \approx |K|_{\mathcal{I}_3} \sum_{j=1}^{k} \omega_j \sqrt{\det\left(\exp\left(\sum_{i=1}^{4} \beta_j^i \log\left(\mathcal{M}_i\right) \right) \right)}.$$

Dihedral angles can be computed using the same methodology.

3.8. Notes

An important part of the above results was first published in Loseille et al. (2010a) and Loseille and Alauzet (2009).

3.8.1. *A short history*

The idea of adapting a mesh is very old. A rather large number of papers have been published dealing with mesh adaptation for numerical simulations, for example, the query "Mesh Adaptation" on Google Scholar found 395,000 results! In most of these works, the adaptation is isotropic and done by splitting elements because of predefined patterns. For instance, a square is split into four squares or a triangle is split into four triangles. The most powerful idea of anisotropic mesh adaptation emerged later, at the end of the 1980s, due to finite element error estimate and mesh generation concerns.

In 1987, Peraire *et al.* proposed a first attempt in 2D by providing error measures involving directions (Peraire et al. 1987). They pointed out the directional property of the interpolation error and the idea to generate elements with aspect ratios. They considered a local mapping procedure to generate elongated elements. They coupled this with an advancing front technique to generate slightly anisotropic meshes, that is, a 1:5 ratio.

Similar approaches have been considered by Löhner (1989) and Selmin and Formaggia (1992). The first attempts in 3D were proposed in the early 1990s by Löhner (1990) and Peraire et al. (1992), but numerical results were almost isotropic and the mesh anisotropy was not clearly visible.

In 1994, Zienkiewicz and Wu (1994) gave a qualified status on the subject. Despite some great success with this new approach, they emphasized that *unfortunately the amount of elongation which can be used in a typical mesh generation by such mapping is small...*

Almost at the same time on the meshing side, Mavriplis (1990) proposed to generate stretched elements with a Delaunay approach in 2D in order to obtain high-aspect ratio triangles in boundary layers and wake regions required by aeronautic numerical simulations. The Delaunay triangulation is performed in a locally stretched space, and the idea of metric has almost emerged.

The year after, George et al. (1991) introduced the use of metrics in a Delaunay mesh generator. They exhibited that the absolute value of the Hessian of the solution field is a metric and they proposed a Delaunay-based mesh generator where edge lengths are computed in a Riemannian metric space. This work generalized all the previous ideas.

The fruitful idea of metric was widely exploited for 2D anisotropic mesh adaptation in the 1990s and even more today. For instance, among many others, see the works of Buscaglia and Dari (1997), Castro-Díaz et al. (1997), Dompierre et al. (1997), Hecht and Mohammadi (1997), and Fortin et al. (1999). Baker (1997) gave the state of the art and wrote: *Mesh generation in three dimensions is difficult enough task in the absence of mesh adaptation and it is only recently that satisfactory three-dimensional mesh generators have become available. [...] Mesh alteration in three dimensions is therefore a rather perilous procedure that should be undertaken with care*. Indeed, 3D meshing is much more complicated as new pathologies occur and the existence of a mesh produced by the mesh generator was not guaranteed. Doing 3D anisotropic mesh adaptation is even more complicated.

These blocking points have been partly solved by proposing a local remeshing approach to adapt the mesh. These approaches start from an existing mesh and modify it to adapt. Therefore, there is no more any existence problem. At the beginning of the 2000s, the first results with "really" 3D anisotropic mesh adaptation for applications were published (Tam et al. 2000; Pain et al. 2001; Bottasso 2004; Belhamadia et al. 2004; Frey and Alauzet 2005; Gruau and Coupez 2005; Li et al. 2005).

In the meantime, new more accurate anisotropic error estimates were proposed: a posteriori estimates (Picasso 2003; Formaggia et al. 2004), a priori estimates (Formaggia and Perotto 2001; Alauzet et al. 2006; Huang 2005) and goal-oriented estimates for functional outputs (Venditti and Darmofal 2003; Jones et al. 2006; Loseille et al. 2010b).

Nowadays, this technology is maturing and impressive results can be obtained. For instance in Alauzet and Mehrenberger (2010b), the authors report a mean anisotropic ratio of 400 and a mean anisotropic quotient of 50,000 (both notions are defined in section 3.5.4 of this chapter) for adapted meshes containing more than 50 million tetrahedra.

Because of its generality, mesh adaptation has been applied to various fields of application and also used with various numerical methods. In all cases, it has brought large improvements in terms of accuracy and CPU performances. Just to give some 3D examples, it has been successfully applied to the sonic boom simulation (Alauzet and Mehrenberger 2010b), multi-fluid flows (Compère et al. 2007; Allain et al. 2009), blast problems (Alauzet et al. 2007), Stefan problems (Belhamadia et al. 2004), metal forming processes (Bruchon et al. 2009), etc. With regard to numerical schemes, mesh adaptation has been coupled with finite volume (Alauzet and Mehrenberger 2010b), finite element (Allain et al. 2009), stabilized finite element (Bruchon et al. 2009), discontinuous finite element (Remacle et al. 2005) methods, etc.

Nowadays, there is a lot of software based on the metric concept. Let us cite `Bamg` (Hecht 1998) and `BL2D` (Laug and Bourochaki 2003) in 2D; `Yams` (Frey 2001) for discrete surface mesh adaptation; and `Feflo.a` (Loseille and Löhner 2009), `Forge3d` (Coupez 2000), `Refine` (Jones et al. 2006), `Gamanic3d` (George 2003), `MAdLib` (Compère et al. 2010), `MeshAdap` (Li et al. 2005), `Mmg3d` (Dobrzynski and Frey 2008), `Mom3d` (Tam et al. 2000), `Tango` (Bottasso 2004) and code from Pain et al. (2001) in 3D. It is worth mentioning that all this software has arisen from different mesh generation methods. The method in George (2003) and Hecht (1998) is based on a global constrained Delaunay kernel. In Laug and Bourochaki (2003), the Delaunay method and the frontal approaches are coupled. Some previous studies (Frey 2001; Loseille and Löhner 2009; Compère et al. 2010; Dobrzynski and Frey 2008) are based on local mesh modifications. And the study by Coupez (2000) is based on the minimal volume principle.

4

Geometric Error Estimate

This chapter addresses the question of how to find the optimal mesh size and anisotropy for the best interpolation a given function. It presents a geometric error estimate based on the P^1 interpolation error. A metric tensor appears naturally from this error estimate. Interpolation errors have been the subject of many studies, in particular for mesh quality purposes (see, for example, Shewchuk 2002; Berzins 1998). From error estimates, abstract error models can be built. The problem of the best-adapted metric can be cast into the optimization of this error model and the optimal metric can be exhibited. The mathematical model also allows the reproduction of some convergence-to-the-limit behaviors, giving some prediction of numerical convergence order for the discrete case. We shall first consider the 1D case and show how the interpolation error can be modeled, and how a calculus of variation produces an optimal metric. The interpolation of a function having a discontinuity is analyzed. Then we shall define a model for the multidimensional case. The error model is derived from an accurate error estimate. Then a definition of the best metric is derived. This time, mesh anisotropy can be taken into account. The convergence order for isotropic and anisotropic adapted meshes is compared in the case of a discontinuous function. Two important options can be distinguished: although often used, *equidistribution*, corresponding to L^∞ error, does not imply a high-order convergence, while the *multiscale* option, corresponding to L^p error, does imply a high-order convergence. We complete these theoretical statements by a series of numerical experiments in order to show examples for which behaviors predicted by the theory are indeed observed.

4.1. The 1D case

4.1.1. *1D metric*

A metric on a given set allows to define the distance between two arbitrary elements of it. We shall call the metric on the interval $[a, b]$ a scalar positive continuous function $\mathcal{M} : x \to \mathcal{M}(x)$ defined on $[a, b]$. It specifies, for any c and d of this interval, the length of segment cd as follows:

$$L_{\mathcal{M}}(cd) = \int_c^d \sqrt{\mathcal{M}}(s)\ ds. \qquad [4.1]$$

Let us consider a mesh of interval $[a, b]$ made of N nodes

$x_0 = a\ <\ x_1, ..., x_i\ <\ x_{i+1}, ..., x_{N-1}\ <\ x_N = b.$

A consequence of the above definition is that a metric can prescribe a particular class of meshes. It is necessary and sufficient that

for any element $[x_i, x_{i+1}]$, *we have* $\int_{x_i}^{x_{i+1}} \sqrt{\mathcal{M}} dx = 1.$

In that case, if we introduce the local continuous mesh size $m_{\mathcal{M}} = \mathcal{M}^{-1/2}$, we have

for any element $[x_i, x_{i+1}]$, $\int_{x_i}^{x_{i+1}} \frac{1}{m_{\mathcal{M}}}\ dx = 1,$

which shows that when $m_{\mathcal{M}}$ is a constant function, it is nothing other than the element size. Another way to view this is to introduce the local continuous node density $d_{\mathcal{M}} = 1/m_{\mathcal{M}}$:

for any interval $[x_i, x_{i+1}]$, *we have* $\int_{x_i}^{x_{i+1}} d_{\mathcal{M}}(x)\ dx = 1.$

It can be verified that the number of nodes (or equivalently intervals) of the mesh is specified by the metric. It is given by

$$C(\mathcal{M}) = \int_a^b \sqrt{\mathcal{M}} dx = \int_a^b 1/m_{\mathcal{M}}(x)\ dx = \int_a^b d_{\mathcal{M}}(x)\ dx. \qquad [4.2]$$

If $C(\mathcal{M})$ is a positive integer, exactly one mesh is described by it; if $C(\mathcal{M})$ is not an integer, no mesh is described by it.

4.1.2. P^1 *Interpolation error bound*

We first recall in short an error bound useful for the sequel. We consider:

– a function u, sufficiently regular, defined on a segment;

– $h = meas([a, b])$, not necessarily small;

– $\Pi_h u$ the $\mathcal{P}_1$ interpolation of u on $[a, b]$: $\Pi_h u$ is an affine function on $[a, b]$, $(\Pi_h u)(a) = u(a)$, $(\Pi_h u)(b) = u(b)$;

– the approximation error defined by $e(x) = u(x) - (\Pi_h u)(x)$.

For any x in $[a, b]$, there exists t_1 and t_2 in $[0, 1]$ such that

$$\begin{aligned} e(a) &= (u - \Pi_h u)(x) + (a - x)(u - \Pi_h u)'(x) + \frac{(a-x)^2}{2} u''(x + t_1(a - x)), \\ e(b) &= (u - \Pi_h u)(x) + (b - x)(u - \Pi_h u)'(x) + \frac{(b-x)^2}{2} u''(x + t_2(b - x)). \end{aligned}$$

Looking for an upper bound of $e = (u - \Pi_h u)$ leads to look for a point x such that $e'(x) = (u - \Pi_h u)'(x) = 0$. After some computation, we get

$$0 = 2(u - \Pi_h u)(x) + \frac{(a-x)^2}{2} u''(x + t_1(a - x)) + \frac{(b-x)^2}{2} u''(x + t_2(b - x))$$

and

$$|(u - \Pi_h u)(x)| \leq \frac{1}{2}(|\frac{(a-x)^2}{2}| + |\frac{(b-x)^2}{2}|) \max_{[a,b]} |u''|.$$

Then

$$|(u - \Pi_h u)(x)| \leq \frac{1}{4} \max_{\xi \in I} \left((a - \xi)^2 + (b - \xi)^2\right) \max_{[a,b]} |u''|.$$

The maximum is reached for $\xi_0 = \frac{(a+b)}{2}$; this implies, $\forall \xi \in I$:

$$|e(\xi)| = |(u - \Pi_h u)(\xi)| \leq \frac{(b-a)^2}{8} \max_{[a,b]} |u''|. \qquad [4.3]$$

This estimate, after it has been modeled in terms of continuous functions, will contribute to the continuous problem statement.

4.1.3. *1D optimal metric*

4.1.3.1. *Optimality condition for norm* $L^p, 1 \leq p < \infty$

The interpolation error vanishes on each vertex. Its maximal value inside each element is estimated by [4.3]. An asymptotic extension would also provide something like the right-hand side of [4.3] as the first term. We propose to represent the interpolation error in the continuous setting by a simplified function inspired from these analyses. This process is a modeling process. It is motivated by the need for a more analyzable mathematical formulation. We assume that the function u is smooth, that its second derivative u'' is everywhere strictly positive and that meshes that we consider are fine enough to allow the second-order term of the interpolation error to represent the whole error well. Let us define the *continuous local* $\mathcal{P}_1$*-interpolation error* as:

$$|e_{\mathcal{M}}(x)| = (d_{\mathcal{M}}(x))^{-2}\, |u''(x)|, \qquad [4.4]$$

where $d_{\mathcal{M}}(x)$ is the node density of the mesh or, equivalently, the inverse local mesh size, that is, the inverse of $m_{\mathcal{M}}(x)$. We want now to find the minimum with respect to the metric $\mathcal{M}$ in a set $\mathcal{U}$ of the L^p norm of the error $e_{\mathcal{M}}$:

$$\min_{\mathcal{M}\in\mathcal{U}} \mathcal{E}_p(\mathcal{M}) \text{ , with } \mathcal{E}_p(\mathcal{M}) = |e_{\mathcal{M}}(x)|^p_{L^p} = \int_a^b (d_{\mathcal{M}}(x)^{-2}|u''(x)|)^p dx. \qquad [4.5]$$

With respect to $d_{\mathcal{M}}$, $\mathcal{U}$ is an open subset of L^2 such that any metric in $\mathcal{U}$ is such that $d_{\mathcal{M}} > 0$ and $d_{\mathcal{M}}^{-2p}$ is of bounded integral. For a node density tending to infinity, the error tends to zero. Let us recall that a sequence of 1D metrics $\mathcal{M}_n$ having n nodes gives a κ*-th order convergence* in L^p norm if the corresponding error satisfies

$$|e_{\mathcal{M}_n}(x)|_{L^p} \;\leq\; const.\; n^{-\kappa}. \qquad [4.6]$$

If we can state such a property for the metric model, this would be a good indication that the κth order convergence is satisfied by meshes built from these metrics. In order to avoid finding the trivial infinitely fine solution, the space of admissible metrics is restricted prescribing the number of nodes:

$$\forall \mathcal{M} \in \mathcal{U},\;\; C(\mathcal{M}) = \bar{C}(d) = \int_a^b d(x)dx \;= N. \qquad [4.7]$$

This gives a linear constraint for variable d. In order to get (at least formally) optimality condition, we can differentiate the functional of [4.5] with respect to d:

$$-2p\int_a^b d^{-2p-1}\left(|u''|\right)^p \delta d\, dx \geq 0,\ \forall \delta d : \int_a^b \delta d\, dx = 0.$$

Thus

$$d_{opt}(x) = K_1.|u''(x)|^{\frac{p}{2p+1}},$$

where K_1 is a constant which we can determine by taking into account the constraint [4.7], we get

$$d_{opt}(x) = \frac{N}{\int |u''|^{\frac{p}{2p+1}} dx'}\, |u''(x)|^{\frac{p}{2p+1}}, \qquad [4.8]$$

or, in terms of the local mesh size,

$$m_{opt}(x) = \frac{\int |u''|^{\frac{p}{2p+1}} dx'}{N}\, |u''(x)|^{\frac{-p}{2p+1}}. \qquad [4.9]$$

The minimum of the functional is written as

$$(\mathcal{E}_p^{opt})^p = \frac{1}{N^{2p}} \left(\int_a^b |u''|^{\frac{p}{2p+1}} dx'\right)^{2p} \left(\int_a^b |u''(x)|^{\frac{1}{2p+1}} dx\right). \qquad [4.10]$$

REMARK 4.1.– The local mesh size, naturally inverse proportional to the number of nodes, is defined by [4.9] only if the second derivative u'' never vanishes. In practice, we replace $|u''|$ by $\max(\varepsilon, |u''|)$, with a small positive ε.

EXAMPLES.–

– For L^1 norm: this is a rather usual norm in image processing, we get

$$m_{opt}^1(x) = \frac{\int |u''|^{1/3} dx'}{N}\, |u''(x)|^{-1/3}. \qquad [4.11]$$

– For L^2 norm, which is the natural option for PDE's, we get

$$m^2_{opt}(x) = \frac{\int |u''|^{2/5} dx'}{N} |u''(x)|^{-2/5}. \qquad [4.12]$$

– For the case of L^∞ norm, due to insufficient smoothness, we cannot get an optimality condition by differentiating the functional. Instead, we can get a formal one by making the power coefficient in [4.9] tend to infinity:

$$m^\infty_{opt}(x) = \frac{\int |u''|^{1/2} dx'}{N} |u''(x)|^{-1/2}. \qquad [4.13]$$

REMARK 4.2.– In the last case, introducing d^∞_{opt},

$$d^\infty_{opt}(x) = \frac{N}{\int |u''|^{\frac{1}{2}} dx'} |u''(x)|^{\frac{1}{2}}. \qquad [4.14]$$

in [4.4] gives a uniform local error,

$$|e^\infty_{\mathcal{M}}(x)| = \frac{(\int |u''|^{\frac{1}{2}} dx')^2}{N^2} \quad \forall x. \qquad [4.15]$$

It is not surprising that the L^∞ norm of error is formally minimum when the local error is uniform. This is an option referred in the literature as the error equidistribution, used in Castro-Díaz et al. (1996) and Habashi et al. (2000).

DEFINITION 4.1.– Approaches minimizing the L^∞ error give a higher impact on local mesh fineness to highest values of u''. They are called *equidistribution methods*, Approaches minimizing a L^p error, with $1 \leq p < \infty$, give also influence to the intermediate values of u''. They are called *multiscale methods*.

4.1.4. *Convergence order of the continuous metric model*

If we take the pth root of expression [4.10], we get

$$\mathcal{E}^{opt}_p = \frac{1}{N^2} \left(\int_a^b |u''|^{\frac{p}{2p+1}} dx' \right)^2 \left(\int_a^b |u''(x)|^{\frac{1}{2p+1}} dx \right)^{1/p}. \qquad [4.16]$$

Since the two integrals are bounded, this shows that the optimal error decreases as the inverse of square of the number of nodes. According to [4.6], this expresses the second-order convergence for $\mathcal{P}_1$ interpolation of the metric sequence obtained by the present adaption strategy. This is not surprising since the function u is assumed to have continuous second derivatives, and the same property also holds for a sequence of uniform meshes.

Let us examine how to look for an optimal metric in the case of a function u having a *discontinuity*. More precisely, u is bounded and smooth on two parts $[a, c]$ and $[c, b]$ of the interval, but is discontinuous at point c with a non-zero step. We choose to represent the $\mathcal{P}_1$ interpolation error as follows:

$$\int_a^b |e_{\mathcal{M}}(x)|^p dx' = \int_a^b (m^2 |u''_\delta(x)|)^p dx,$$

$$\text{with} \quad u''_\delta(x) \;=\; \delta^{-2}(u(x+\delta) - 2u(x) + u(x-\delta)). \qquad [4.17]$$

In the sequel, δ is assumed to be smaller than mesh size m. Due to the regularity of u in the complement of the discontinuity location, the function u''_δ is bounded except on an interval of size 3δ where

$$|u''_\delta(x)| \;\leq\; 4\,\delta^{-2} \,\max |u|.$$

Using this regularized error model is justified by the following remarks:

– $u''_\delta \to \frac{\partial^2 u}{\partial x^2}$ in space $\mathcal{D}'$ of distributions[1] when $\delta \to 0$;

– on the element $[x_i, x_{i+1}]$ containing the discontinuity, the interpolation error $\int_{x_i}^{x_{i+1}} |\Pi_h u - u|^p dx$ is asymptotically smaller than $\int_{x_i}^{x_{i+1}} meas([x_i, x_{i+1}])^{2p} \, |u''_\delta|^p dx$.

A *first* consequence in using this regularization is that in the case where δ is equal to the local mesh size, this approximative Hessian will specify close to the discontinuity a mesh density in accordance to the discontinuity capturing densities identified in lemma 2.2 of Chapter 2. *Second*, we remark that for γ such that $0 \;<\; \gamma \;\leq\; 1/2$,

$$||u''_\delta||_{L^\gamma} \;\text{ is bounded independantly of }\; \delta \qquad [4.18]$$

(and not bounded for $\gamma \;>\; 1/2$). The resulting optimal error is written as

$$\mathcal{E}_p^{opt} \;=\; \frac{1}{N^2} \left(\int_a^b |u''_\delta|^{\frac{p}{2p+1}} dx' \right)^2 \left(\int_a^b |u''_\delta(x)|^{\frac{1}{2p+1}} dx \right)^{1/p}, \qquad [4.19]$$

1 In fact, in the space $L^{\frac{1}{2}}$.

then for $p \leq \frac{5}{2}$, we have

$$\mathcal{E}_p^{opt} \leq \frac{K_2}{N^2},$$

where K_2 is a bounded constant, due to [4.18]. We have demonstrated the first part of the following lemma:

LEMMA 4.1.– For $p \leq \frac{5}{2}$, and in particular for $p = 2$, multiscale methods can produce higher order L^p convergence on discontinuities, while equidistribution methods cannot.

Concerning the second part of proof, we recall that continuous functions cannot converge in L^∞ to a discontinuous function. □

4.2. Discrete-continuous duality for linear interpolation error

Now we aim at applying the framework of previous chapter in the context of error estimation in 2D and 3D. However, as our intent is to propose a fully discrete-continuous duality, it is not sufficient to derive only the optimal mesh arising from an interpolation error bound as in classical studies on interpolation error (Castro-Díaz et al. 1997; Frey and Alauzet 2005; Huang 2005). Instead, we want to evaluate the interpolation error for any functions on any continuous meshes without imposing some optimality conditions such as alignment, equidistribution, and so on.

Let $(\mathcal{M}(\mathbf{x}))_{\mathbf{x}\in\Omega}$ be a continuous mesh of a domain Ω, and let u be a nonlinear function, which is assumed to be only twice continuously differentiable. For any triangular or tetrahedral mesh $\mathcal{H}$ built on an approximation Ω_h of Ω, we can define the linear continuous interpolation as:

DEFINITION 4.2.– The linear interpolation operator $\Pi_{\mathcal{H}}$ maps any continuous function φ defined on Ω_h to a continuous function φ_h which is affine on any element of Ω_h and equal to φ on any vertex:

$$\Pi_{\mathcal{H}}\varphi(\mathbf{x}) = \varphi(\mathbf{x}) \ \forall\ \mathbf{x} \text{ vertex of } \mathcal{H} \text{ and } \Pi_{\mathcal{H}}\varphi|_T \text{ is affine } \forall\ T \text{ element of } \mathcal{H}.$$

We seek a well-posed definition of the continuous linear interpolation error $\|u - \pi_{\mathcal{M}}u\|_{\mathbf{L}^1(\Omega)}$ related to a continuous mesh $(\mathcal{M}(\mathbf{x}))_{\mathbf{x}\in\Omega}$, which implies a well-posed definition of a linear continuous interpolate $\pi_{\mathcal{M}}u$. More precisely, we would like the continuous linear interpolation error to be a reliable mathematical model of $\|u - \Pi_h u\|_{\mathbf{L}^1(\Omega_h)}$, where Π_h is defined by a mesh $\mathcal{H}$ of a discretized domain Ω_h, which is a unit mesh with respect to $(\mathcal{M}(\mathbf{x}))_{\mathbf{x}\in\Omega}$. *In fine*, this means that considering $\|u - \pi_{\mathcal{M}} u\|_{\mathbf{L}^1(\Omega)}$ will be equivalent to consider $\|u - \Pi_h u\|_{\mathbf{L}^1(\Omega_h)}$.

The error analysis is first done locally, that is, in a tangent space of $(\mathcal{M}(\mathbf{x}))_{\mathbf{x}\in\Omega}$ at a given point $\mathbf{a}$. In the tangent space, $(\mathcal{M}(\mathbf{x}))_{\mathbf{x}\in\Omega}$ reduces to the continuous element $\mathcal{M}(\mathbf{a})$. In other words, the analysis is performed locally at the element level. The function u is approximated by its local quadratic Taylor expansion. Indeed, terms of order greater than two can be neglected while studying the linear interpolation error. Then a global error estimate is derived by taking into account the variation of the continuous mesh and of the function.

4.2.1. *Interpolation error in* L^1 *norm for quadratic functions*

In this section, we consider a quadratic function u defined on a domain $\Omega \subset \mathbb{R}^3$. The function is given by its matrix representation

$$\forall \mathbf{x} \in \Omega, \quad u(\mathbf{x}) = \frac{1}{2}\, {}^t\mathbf{x}\, H\, \mathbf{x},$$

where H is a symmetric matrix representing the Hessian of u. For every symmetric matrix H, $|H|$ denotes the positive symmetric matrix deduced from H by taking the absolute values of its eigenvalues. The function u is linearly interpolated on a tetrahedron K defined by its vertices list: $K = (\mathbf{x}_1, \mathbf{x}_2, \mathbf{x}_3, \mathbf{x}_4)$. $|K|$ denotes the Euclidean volume of K. We denote by $\Pi_h u$ the linear interpolate of u on K. We can now state the following result.

PROPOSITION 4.1.– For every quadratic function u, its linear interpolation error in $\mathbf{L}^1$ norm on a tetrahedron K verifies

$$\|u - \Pi_h u\|_{\mathbf{L}^1(K)} \leq \frac{|K|}{40} \sum_{i=1}^{6} {}^t\mathbf{e}_i |H| \mathbf{e}_i,$$

where $(\mathbf{e}_i)_{i=1,6}$ is the set of edges of K.

PROOF.– The proof consists of deriving an exact error estimate of the pointwise interpolation error within the element K: $e(\mathbf{x}) = (u - \Pi_h u)(\mathbf{x})$ for $\mathbf{x} \in K$. This error is then integrated over K. To derive e, we use the standard reference element technique. Reference element K_{ref} is defined by the coordinates of its four vertices:

$$\widehat{\mathbf{x}}_1 = {}^t(0,0,0),\ \widehat{\mathbf{x}}_2 = {}^t(1,0,0),\ \widehat{\mathbf{x}}_3 = {}^t(0,1,0),\ \text{and}\ \widehat{\mathbf{x}}_4 = {}^t(0,0,1).$$

All the computations are done on K_{ref}, and the result is then mapped onto the current element K by using the following affine mapping:

$$\mathbf{x} = \mathbf{v}_1 + B_K \,\widehat{\mathbf{x}} \text{ with } B_K = [\mathbf{x}_2 - \mathbf{x}_1, \mathbf{x}_3 - \mathbf{x}_1, \mathbf{x}_4 - \mathbf{x}_1], \;\; \mathbf{x} \in K, \;\; \hat{\mathbf{x}} \in K_{ref}.$$

The matrix B_k can be expressed as a function of the following edges:

$$\mathbf{e}_1 = \mathbf{x}_2 - \mathbf{x}_1, \;\; \mathbf{e}_2 = \mathbf{x}_3 - \mathbf{x}_1, \quad \text{and} \quad \mathbf{e}_3 = \mathbf{x}_4 - \mathbf{x}_1,$$

so that $B_K = [\mathbf{e}_1, \mathbf{e}_2, \mathbf{e}_3]$. The quadratic function u reads in the frame of K_{ref}:

$$u(\mathbf{x}(\widehat{\mathbf{x}})) = \frac{1}{2}\,{}^t\mathbf{x}_1\, H\, \mathbf{x}_1 + \frac{1}{2}\,{}^t\mathbf{x}_1\, H\, B_K\, \widehat{\mathbf{x}} + \frac{1}{2}\,{}^t\widehat{\mathbf{x}}\,{}^tB_K\, H\, \mathbf{x}_1 + \frac{1}{2}\,{}^t\widehat{\mathbf{x}}\,{}^tB_K\, H\, B_K\, \widehat{\mathbf{x}}.$$

As we consider the linear interpolation, linear and constant terms of $u(\mathbf{x}(\widehat{\mathbf{x}}))$ are exactly interpolated. Without loss of generality, these terms are neglected and only quadratic terms are kept. Indeed, if we consider $\tilde{u}(\mathbf{x}) = \frac{1}{2}\,{}^t\widehat{\mathbf{x}}\,{}^tB_K\, H\, B_K\, \widehat{\mathbf{x}}$, then we get

$$e(\mathbf{x}) = (u - \Pi_h u)(\mathbf{x}) = (\tilde{u} - \Pi_h \tilde{u})(\mathbf{x}).$$

We can now consider $\tilde{u}$ instead of u. However, for the sake of clarity, we keep on writing u and not $\tilde{u}$. We rewrite u in a matrix form:

$$u(\mathbf{x}(\widehat{\mathbf{x}})) = \frac{1}{2}{}^t\begin{pmatrix} \hat{x} \\ \hat{y} \\ \hat{z} \end{pmatrix} \begin{bmatrix} {}^t\mathbf{e}_1 H \mathbf{e}_1 & {}^t\mathbf{e}_1 H \mathbf{e}_2 & {}^t\mathbf{e}_1 H \mathbf{e}_3 \\ {}^t\mathbf{e}_2 H \mathbf{e}_1 & {}^t\mathbf{e}_2 H \mathbf{e}_2 & {}^t\mathbf{e}_2 H \mathbf{e}_3 \\ {}^t\mathbf{e}_3 H \mathbf{e}_1 & {}^t\mathbf{e}_3 H \mathbf{e}_2 & {}^t\mathbf{e}_3 H \mathbf{e}_3 \end{bmatrix} \begin{pmatrix} \hat{x} \\ \hat{y} \\ \hat{z} \end{pmatrix}.$$

u in K_{ref} reads

$$\begin{aligned} u(\mathbf{x}(\widehat{\mathbf{x}})) = \frac{1}{2} \;& \big[({}^t\mathbf{e}_1 H \mathbf{e}_1)\,\hat{x}^2 \;+\; ({}^t\mathbf{e}_2 H \mathbf{e}_2)\,\hat{y}^2 \;+\; ({}^t\mathbf{e}_3 H \mathbf{e}_3)\,\hat{z}^2 \\ & + 2({}^t\mathbf{e}_1 H \mathbf{e}_2)\,\hat{x}\hat{y} \;+\; 2({}^t\mathbf{e}_1 H \mathbf{e}_3)\,\hat{x}\hat{z} \;+\; 2({}^t\mathbf{e}_2 H \mathbf{e}_3)\,\hat{y}\hat{z}\big]. \end{aligned}$$

u is now linearly interpolated on K_{ref}. Its linear interpolate $\Pi_h u(\widehat{\mathbf{x}})$ is written as $a\widehat{x} + b\widehat{y} + c\widehat{z} + d$, where coefficients $(a, b, c, d) \in \mathbb{R}^4$ satisfy the following linear system, ensuring the $\mathcal{P}_1$ exactness, that is, $\Pi_h u(\mathbf{v}_i) = u(\mathbf{v}_i)$ for all $i \in [1, 4]$:

$$\begin{cases} \Pi_h u(\mathbf{v}_1) = d = u(\mathbf{x}(0,0,0)) = 0, \\ \Pi_h u(\mathbf{v}_2) = a = u(\mathbf{x}(1,0,0)) = \frac{1}{2}\left({}^t\mathbf{e}_1 H \mathbf{e}_1\right), \\ \Pi_h u(\mathbf{v}_3) = b = u(\mathbf{x}(0,1,0)) = \frac{1}{2}\left({}^t\mathbf{e}_2 H \mathbf{e}_2\right), \\ \Pi_h u(\mathbf{v}_4) = c = u(\mathbf{x}(0,0,1)) = \frac{1}{2}\left({}^t\mathbf{e}_3 H \mathbf{e}_3\right). \end{cases}$$

The solution of the previous linear system gives the final expression of $\Pi_h u$:

$$\Pi_h u(\mathbf{x}(\widehat{\mathbf{x}})) = \frac{1}{2}\left[\left({}^t\mathbf{e}_1 H \mathbf{e}_1\right)\hat{x} + \left({}^t\mathbf{e}_2 H \mathbf{e}_2\right)\hat{y} + \left({}^t\mathbf{e}_3 H \mathbf{e}_3\right)\hat{z}\right].$$

The exact pointwise interpolation error $e(\mathbf{x})$ is then given by

$$\begin{aligned} e(\mathbf{x}(\widehat{\mathbf{x}})) = &\frac{1}{2}[\left({}^t\mathbf{e}_1 H \mathbf{e}_1\right)(\hat{x}^2 - \hat{x}) + \left({}^t\mathbf{e}_2 H \mathbf{e}_2\right)(\hat{y}^2 - \hat{y}) + \left({}^t\mathbf{e}_3 H \mathbf{e}_3\right)(\hat{z}^2 - \hat{z}) \\ &+ 2\left({}^t\mathbf{e}_1 H \mathbf{e}_2\right)\hat{x}\hat{y} + 2\left({}^t\mathbf{e}_1 H \mathbf{e}_3\right)\hat{x}\hat{z} + 2\left({}^t\mathbf{e}_2 H \mathbf{e}_3\right)\hat{y}\hat{z}]. \end{aligned}$$

From this equality, an estimate in $\mathbf{L}^1$, $\mathbf{L}^2$, or $\mathbf{H}^1$ can be deduced by considering the change of variables given by the mapping B_K. Indeed, for every function F, its integration over K can be computed through its expression in K_{ref}:

$$\int_K F(\mathbf{x})\,\mathrm{d}x\mathrm{d}y\mathrm{d}z = \int_{K_{ref}} F(\mathbf{x}(\widehat{\mathbf{x}}))\,|\det(B_K)|\,\mathrm{d}\hat{x}\mathrm{d}\hat{y}\mathrm{d}\hat{z}.$$

As $6\,|K| = \det(B_k)$, the previous equality becomes

$$\int_K F(\mathbf{x})\,\mathrm{d}x\mathrm{d}y\mathrm{d}z = 6|K| \int_{K_{ref}} F(\mathbf{x}(\widehat{\mathbf{x}}))\,\mathrm{d}\hat{x}\mathrm{d}\hat{y}\mathrm{d}\hat{z}.$$

Consequently, the interpolation error in $\mathbf{L}^1$ norm is evaluated by a direct integration of $|e(\mathbf{x})|$. When u is concave or convex, we have the following: $|(u - \Pi_h u)(\mathbf{x})| = (u - \Pi_h u)(\mathbf{x})$ in the convex case and $|(u - \Pi_h u)(\mathbf{x})| = -(u - \Pi_h u)(\mathbf{x})$ in the concave case. The error reads

$$\|u - \Pi_h u\|_{\mathbf{L}^1(K)} = \frac{|K|}{40}\Big|2({}^t\mathbf{e}_1\, H\, \mathbf{e}_2 + {}^t\mathbf{e}_1\, H\, \mathbf{e}_3 + {}^t\mathbf{e}_2\, H\, \mathbf{e}_3) \\ -3({}^t\mathbf{e}_1\, H\, \mathbf{e}_1 + {}^t\mathbf{e}_2\, H\, \mathbf{e}_2 + {}^t\mathbf{e}_3\, H\, \mathbf{e}_3)\Big|.$$

Using the conventions of Figure 3.11, the crossed terms can be expressed only in terms of $\mathbf{e}_i\, H\, \mathbf{e}_i$ for $i = 1, \ldots, 6$:

$$\begin{cases} 2\,{}^t\mathbf{e}_1\, H\, \mathbf{e}_2 = {}^t\mathbf{e}_1\, H\, \mathbf{e}_1 + {}^t\mathbf{e}_2\, H\, \mathbf{e}_2 - {}^t\mathbf{e}_4\, H\, \mathbf{e}_4, \\ 2\,{}^t\mathbf{e}_1\, H\, \mathbf{e}_3 = {}^t\mathbf{e}_1\, H\, \mathbf{e}_1 + {}^t\mathbf{e}_3\, H\, \mathbf{e}_3 - {}^t\mathbf{e}_5\, H\, \mathbf{e}_5, \\ 2\,{}^t\mathbf{e}_2\, H\, \mathbf{e}_3 = {}^t\mathbf{e}_2\, H\, \mathbf{e}_2 + {}^t\mathbf{e}_3\, H\, \mathbf{e}_3 - {}^t\mathbf{e}_6\, H\, \mathbf{e}_6. \end{cases}$$

We deduce

$$\left|2({}^t\mathbf{e}_1\, H\, \mathbf{e}_2 + {}^t\mathbf{e}_1\, H\, \mathbf{e}_3 + {}^t\mathbf{e}_2\, H\, \mathbf{e}_3) - 3({}^t\mathbf{e}_1\, H\, \mathbf{e}_1 + {}^t\mathbf{e}_2\, H\, \mathbf{e}_2 + {}^t\mathbf{e}_3\, H\, \mathbf{e}_3)\right| \\ = \left|\sum_{i=1}^{6} {}^t\mathbf{e}_i\, H\, \mathbf{e}_i\right|. \square$$

The same proof applies in 2D. For a quadratic function u, the linear interpolation error on a triangle K is given by

$$\|u - \Pi_h u\|_{\mathbf{L}^1(K)} \leq \frac{|K|}{24} \sum_{i=1}^{3} {}^t\mathbf{e}_i |H| \mathbf{e}_i.$$

Error estimates in the $\mathbf{L}^2$ norm and in the $\mathbf{H}^1$ norm can also be derived. We refer to Berzins (1998); Nadler (1985) and references therein for their evaluations. These error estimates are classically used to exhibit mesh quality functions and to obtain the best element shape minimizing the interpolation error (Bank and Smith 1997). In the continuous mesh framework, the interpolation error estimate in the $\mathbf{L}^1$ norm is used to prove some exactness properties of the continuous linear interpolate.

4.2.2. *Linear interpolation on a continuous element*

Let $\mathcal{M}$ be a continuous element and u be a quadratic function with positive Hessian. We study the interpolation error for the class of all unit discrete elements with respect to $\mathcal{M}$, given by definition 3.8. Figure 3.9 depicts for a given metric tensor $\mathcal{M}$ some unit elements. We can now state the main result.

THEOREM 4.1.– For all unit elements K with respect to $\mathcal{M}$, the interpolation error of u in the $\mathbf{L}^1$ norm does not depend on the element shape and is only a function of the Hessian H of u and continuous element $\mathcal{M}$.

– In 3D, for all unit elements K for $\mathcal{M}$, the following equality holds:

$$\boxed{\|u - \Pi_h u\|_{\mathbf{L}^1(K)} = \frac{\sqrt{2}}{240}\, \det(\mathcal{M}^{-\frac{1}{2}})\, \mathit{trace}(\mathcal{M}^{-\frac{1}{2}}\, H\, \mathcal{M}^{-\frac{1}{2}}).} \qquad [4.20]$$

– In 2D, for all unit elements K for $\mathcal{M}$, the following equality holds:

$$\boxed{\|u - \Pi_h u\|_{\mathbf{L}^1(K)} = \frac{\sqrt{3}}{64}\, \det(\mathcal{M}^{-\frac{1}{2}})\, \mathit{trace}(\mathcal{M}^{-\frac{1}{2}}\, H\, \mathcal{M}^{-\frac{1}{2}}).}$$

PROOF.– According to proposition 4.1, the interpolation error in the $\mathbf{L}^1$ norm of a quadratic positive function u on an element K is

$$\|u - \Pi_h u\|_{\mathbf{L}^1(K)} = \frac{|K|}{40} \sum_{i=1}^{6} {}^t\mathbf{e}_i H \mathbf{e}_i.$$

Then, if K is unit with respect to $\mathcal{M}$, the previous interpolation error is expressed by

$$\|u - \Pi_h u\|_{\mathbf{L}^1(K)} = \frac{\sqrt{2}}{240}\, \det(\mathcal{M}^{-\frac{1}{2}})\, \mathrm{trace}(\mathcal{M}^{-\frac{1}{2}}\, H\, \mathcal{M}^{-\frac{1}{2}}),$$

because of the geometric invariants related to the volume, relation [3.10], and to the square lengths of the edges, relation [3.11]. □

We note the strong analogy with classical interpolation error estimate for Lagrange interpolation (Ciarlet 1978):

– The term $\det(\mathcal{M}^{-\frac{1}{2}})$ stands for the Jacobian of the affine transformation from the reference element $\hat{K}$ onto the current element K. In our continuous framework, it is the Jacobian of the affine mapping from the reference continuous element unit ball $\mathcal{B}_{I_3}$ onto the current continuous element unit ball $\mathcal{B}_{\mathcal{M}}$.

– The term trace$(\mathcal{M}^{-\frac{1}{2}} H \mathcal{M}^{-\frac{1}{2}})$ stands for the seminorm involved in classical error estimates. Generally, this seminorm contains the anisotropic behavior of the estimate. In the continuous framework, the trace-term gives the alignment correlation between the principal directions of the Hessian H and the principal directions of the metric $\mathcal{M}$.

Relation [4.20] shows that the infinite set of discrete elements that are unit for a given continuous element $\mathcal{M}$ achieves the same interpolation error and moreover shows that this interpolation error is only expressed with continuous quantities: the continuous element $\mathcal{M}$ and the Hessian of the function u. Consequently, theorem 4.1 points out that the metric alone contains enough information to describe completely the linear interpolation error in the $\mathbf{L}^1$ norm. In other words, this theorem confirms that the use of metric-based mesh adaptation is particularly well suited to control anisotropically the interpolation error. In the past, this efficiency has been observed practically on real-life problems (see, for instance, Dompierre et al. 1997; Tam et al. 2000; Pain et al. 2001; Bottasso 2004; Formaggia et al. 2004; Schall et al. 2004; Frey and Alauzet 2005; Loseille et al. 2007).

4.2.3. *Continuous linear interpolate*

The main difficulty in defining the continuous linear interpolate is to connect a discrete error computed on an element to a local continuous error that is defined pointwise. Indeed, the discrete interpolation error in norm $\mathbf{L}^1$ is defined on each element K. On the contrary, a continuous mesh is a function $\mathbf{x} \mapsto \mathcal{M}(\mathbf{x})$ defined at each point $\mathbf{x}$ of Ω. Suppose now that the continuous mesh $(\mathcal{M}(\mathbf{x}))_{\mathbf{x}\in\Omega}$ is varying and that the function u is no more quadratic but only twice continuously differentiable. While equality [4.20] of theorem 4.1 does not hold anymore, all the terms of the right-hand side $\mathcal{M}$ and H are well-defined continuously. The definition of a continuous interpolate follows up from this consideration. We denote by u_Q the following quadratic approximation of a smooth function u: at point $\mathbf{a}$, u_Q is defined in the vicinity of $\mathbf{a}$ as the truncated second-order Taylor expansion of u:

$$\forall \mathbf{x} \in \mathcal{V}(\mathbf{a}), \;\; u_Q(\mathbf{a};\mathbf{x}) = u(\mathbf{a}) + \nabla u(\mathbf{a})(\mathbf{x}-\mathbf{a}) + \frac{1}{2}\langle(\mathbf{x}-\mathbf{a}),\, H(\mathbf{a})(\mathbf{x}-\mathbf{a})\rangle.$$

When no confusion is possible, the notation $u_Q(\mathbf{a};\mathbf{x})$ is replaced by $u_Q(\mathbf{x})$. u_Q is a complete quadratic form composed of a constant term, a linear term and finally a quadratic term. The first result of this section provides an equivalence formula between discrete and continuous views locally around a point $\mathbf{a}$ of the domain. In the vicinity of $\mathbf{a}$, u_Q approximates u and $(\mathcal{M}(\mathbf{x}))_{\mathbf{x}\in\Omega}$ reduces to $\mathcal{M}(\mathbf{a})$ in the tangent space. We can now state the main result.

THEOREM 4.2 (discrete-continuous equivalence).– Let u be a twice continuously differentiable function of a domain Ω and $(\mathcal{M}(\mathbf{x}))_{\mathbf{x}\in\Omega}$ be a continuous mesh of Ω. Then there exists a unique continuous linear interpolate function $\pi_{\mathcal{M}}$ such that

$$\forall \mathbf{a} \in \Omega, \quad |u - \pi_{\mathcal{M}} u|(\mathbf{a}) = 2\,\frac{\|u_Q - \Pi_h u_Q\|_{\mathbf{L}^1(K)}}{|K|}$$

for every K unit element with respect to $\mathcal{M}(\mathbf{a})$.

The proof is given hereafter; we first look at the consequences of this theorem.

COROLLARY 4.1 (Continuous linear interpolate).– Let u be a twice continuously differentiable function of a domain Ω and $(\mathcal{M}(\mathbf{x}))_{\mathbf{x}\in\Omega}$ be a continuous mesh of Ω. Then the continuous linear interpolate satisfies the estimate in 3D:

$$\begin{aligned}\forall \mathbf{a} \in \Omega, \quad |u - \pi_{\mathcal{M}} u|(\mathbf{a}) &= \frac{1}{10}\,\mathrm{trace}\big(\mathcal{M}(\mathbf{a})^{-\frac{1}{2}}\,|H(\mathbf{a})|\,\mathcal{M}(\mathbf{a})^{-\frac{1}{2}}\big)\\ &= \frac{1}{10}\left(d(\mathbf{a})^{-\frac{2}{3}}\sum_{i=1}^{3} r_i(\mathbf{a})^{\frac{2}{3}}\ {}^t\mathbf{v}_i(\mathbf{a})\,|H(\mathbf{a})|\,\mathbf{v}_i(\mathbf{a})\right).\end{aligned}$$

In 2D, the estimate is

$$\begin{aligned}\forall \mathbf{a} \in \Omega, \quad |u - \pi_{\mathcal{M}} u|(\mathbf{a}) &= \frac{1}{8}\,\mathrm{trace}\big(\mathcal{M}(\mathbf{a})^{-\frac{1}{2}}\,|H(\mathbf{a})|\,\mathcal{M}(\mathbf{a})^{-\frac{1}{2}}\big)\\ &= \frac{1}{8}\left(d(\mathbf{a})^{-1}\sum_{i=1}^{2} r_i(\mathbf{a})\ {}^t\mathbf{v}_i(\mathbf{a})\,|H(\mathbf{a})|\,\mathbf{v}_i(\mathbf{a})\right)\end{aligned}$$

(in short for both dimensions: $|u - \pi_{\mathcal{M}} u|(\mathbf{a}) = \frac{1}{2(d+2)}\,\mathrm{trace}\big(\mathcal{M}(\mathbf{a})^{-\frac{1}{2}}|H(\mathbf{a})|\,\mathcal{M}(\mathbf{a})^{-\frac{1}{2}}\big)$).

PROOF OF COROLLARY.– In 3D, for all unit elements K with respect to $\mathcal{M}(\mathbf{a})$, the error estimation [4.20] can be rewritten as follows for the quadratic function u_Q approximating u in the vicinity of $\mathbf{a}$:

$$\frac{\|u_Q - \Pi_h u_Q\|_{\mathbf{L}^1(K)}}{|K|} = \frac{1}{20}\,\mathrm{trace}(\mathcal{M}(\mathbf{a})^{-\frac{1}{2}}\,|H(\mathbf{a})|\,\mathcal{M}(\mathbf{a})^{-\frac{1}{2}}).$$

Then, expressing $\mathcal{M}(\mathbf{a})$ as a function of the continuous mesh parameters given by the decomposition of proposition 3.2 leads to

$$\frac{\|u_Q - \Pi_h u_Q\|_{\mathbf{L}^1(K)}}{|K|} = \frac{1}{20}\left(d(\mathbf{a})^{-\frac{2}{3}} \sum_{i=1}^{3} r_i(\mathbf{a})^{\frac{2}{3}}\, {}^t\mathbf{v}_i(\mathbf{a})\, |H(\mathbf{a})|\, \mathbf{v}_i(\mathbf{a}) \right),$$

where the $(\mathbf{v}_i(a))_{i=1,3}$ stand for the eigenvectors of $\mathcal{M}(\mathbf{a})$. □

This result shows that the continuous pointwise linear interpolation can be decomposed into the product of two terms:

– a first term that controls the accuracy; this density term is directly connected to the size of the continuous element;

– a second term that measures alignment deviation between the continuous element orientation and the anisotropy features of the function u.

REMARK 4.3 (Convergence order).– Note that if the first term (involving the density) is rewritten in terms of metric size accordingly to proposition 3.2, $d(\mathbf{a})^{-\frac{2}{3}}$ in 3D and $d(\mathbf{a})^{-1}$ in 2D are both of the order of $h^{-2}(\mathbf{a})$. This shows the typical second order of convergence for the continuous linear interpolation error. □

REMARK 4.4 (Further terms).– The metric density $d_{\mathcal{M}}$ is generally a smooth function representing the discrete vertex density $d_{\mathcal{H}}$ of the unit mesh $\mathcal{H}$ with respect to $\mathbf{M}$. In contrast, the density $d_{\mathcal{H}}$ is a sum of Dirac located at vertices and can be therefore considered as highly oscillating. The metric density is thus a *homogenization* of the unit mesh density. To illustrate this, a continuous mesh convergent sequence can be defined according to the continuous mesh complexity $\mathcal{C}(\mathcal{M}) = N$ and a reference metric $\mathcal{M}_1$ of complexity equal to one $\mathcal{C}(\mathcal{M}_1) = 1$:

$$\mathcal{M}_N = \mathcal{C}(\mathcal{M})\, \mathcal{M}_1 = N\, \mathcal{M}_1, \;\; N \to \infty.$$

We can consider an average mesh size h corresponding to complexity N, and reformulate the continuous mesh convergent sequence as:

$$\mathcal{M}_h = h^{-d} \mathcal{M}_1, \;\; h \to 0,$$

where d is the domain dimension. The continuous and discrete densities are close to each other in the sense that:

$$\forall \phi \in \mathcal{D}(\Omega),\; (d_{\mathcal{M}_N}, \phi) - (d_{\mathcal{H}_N}, \phi) \to 0 \textbf{ as } N \to \infty,$$

where $\mathcal{H}_N$ is a unit mesh for $\mathcal{M}_N$, and $\mathcal{D}(\Omega)$ is the subset of $\mathcal{C}^\infty(\bar{\Omega})$ functions with compact support. Similarly, given a smooth function u defined on Ω, and its P^1 interpolation $\Pi_h u$ on unit mesh $\mathcal{H}$ for $\mathcal{M}_h$, the interpolation error $u - \Pi_h u$ is also a highly oscillatory function. Now, according to Loseille and Alauzet (2011a,b), it is possible to homogenize the discrete interpolation error using the *continuous interpolation error* of theorem 4.2 expressed in terms of the Hessian H_u of u in such a way that, for $h \to 0$,

$$|u - \Pi_h u| = h^2 e^{\mathbf{homog}}_{interp}(u) + h^2 e^{oscill}_{interp}(u) + h^2\ o(h)_{L^2(\Omega)}$$
$$\text{with}\quad e^{\mathbf{homog}}_{interp}(u) = |u - \pi_{\mathcal{M}_1} u|,$$

where $e^{\mathbf{homog}}_{interp}(u)$ is the homogenized interpolation error, $h^2\ o(h)_{L^2(\Omega)}$ is an error term of order strictly higher than two according to L^2 analysis and $h^2\ e^{oscill}_{interp}(u)$ is the oscillatory component of the interpolation error, which is also a error term of order strictly higher than two since $e^{oscill}_{interp} \to 0\ in\ \mathcal{D}'(\Omega)$. The goal of our analysis is to propose models for $e^{\mathbf{homog}}_{interp}(u)$, considering the sequence $\mathcal{M}_h$ when necessary. □

It is possible to give a geometric interpretation of the estimate of the above corollary. This interpretation illustrates the impact of a continuous mesh on the error iso-values and, consequently, gives some clue toward the control of the error by means of a continuous mesh. The term $\mathcal{M}^{-\frac{1}{2}}(\mathrm{a})\,|H(\mathrm{a})|\,\mathcal{M}^{-\frac{1}{2}}(\mathrm{a})$ corresponds to the frame change related to the continuous mesh local orientation. Given a symmetric matrix $|H(\mathrm{a})|$, the corresponding quadratic form is $f = \frac{1}{2}\,{}^t\mathrm{x}\,|H(\mathrm{a})|\,\mathrm{x}$.

The matrix $\mathcal{M}^{-\frac{1}{2}}(\mathrm{a})\,|H(\mathrm{a})|\,\mathcal{M}^{-\frac{1}{2}}(\mathrm{a})$ corresponds to a new quadratic form observed in the space deformed by $\mathcal{M}(\mathrm{a})$. Indeed, if we consider the change of coordinates $\tilde{\mathrm{x}} = \mathcal{M}^{\frac{1}{2}}(\mathrm{a})\,\mathrm{x}$, we define a new quadratic form $\tilde{f}$:

$$\begin{aligned}\tilde{f}(\tilde{\mathrm{x}}) &= {}^t\mathrm{x}\,|H(\mathrm{a})|\,\mathrm{x} = {}^t(\mathcal{M}^{-\frac{1}{2}}(\mathrm{a})\,\tilde{\mathrm{x}})\,|H(\mathrm{a})|\,\mathcal{M}^{-\frac{1}{2}}(\mathrm{a})\,\tilde{\mathrm{x}}\\ &= {}^t\tilde{\mathrm{x}}\left(\mathcal{M}^{-\frac{1}{2}}(\mathrm{a})\,|H(\mathrm{a})|\,\mathcal{M}^{-\frac{1}{2}}(\mathrm{a})\right)\tilde{\mathrm{x}}.\end{aligned}$$

Iso-values of $\tilde{f}$ and f are different when they are seen in the canonical Euclidean space. In fact, viewing a quadratic form for different continuous meshes changes its iso-values in the Euclidean space, that is, the real physical space. It is then possible to control the error iso-values by modifying $\mathcal{M}$. This is the main principle of mesh adaptation, but here formulated in a continuous framework. The mesh adaptation principle is as follows: *Classical metric-based mesh adaptation consists of finding a metric field that provides isotropic iso-values of the error function*; see the pioneer work in Castro-Díaz et al. (1997). These results demonstrate that both the interpolation error and the linear interpolate Π_h have continuous counterparts. It is then a step forward in finding a complete duality between the discrete and the continuous views. From a practical point of view, we deduce the following analogy. Given a unit mesh $\mathcal{H}$ of a domain Ω_h with respect to a continuous mesh $(\mathcal{M}(\mathbf{x}))_{\mathbf{x}\in\Omega}$, the global interpolation error is

$$\|u - \Pi_h u\|_{\mathbf{L}^1(\Omega_h)} = \sum_{K\in\mathcal{H}} \|u - \Pi_h u\|_{\mathbf{L}^1(K)}. \qquad [4.21]$$

In the continuous case, the discrete summation becomes an integral:

$$\|u - \pi_{\mathcal{M}} u\|_{\mathbf{L}^1(\Omega)} = \int_\Omega |u - \pi_{\mathcal{M}} u|(\mathbf{x})\,\mathrm{d}\mathbf{x}. \qquad [4.22]$$

Note that there is no global guarantee on the continuous interpolation error reliability given by relation [4.22]. For instance, there is no *a priori* relationship between [4.21] and [4.22]. The only guarantee is the local equivalence given by theorem 4.2. However, the local guarantee becomes global when the mesh is unit with respect to a constant metric tensor (this does not necessary imply that the mesh is uniform) and when the function is quadratic. In this specific case, by neglecting error due to the boundary discretization, we have the equality

$$2\,\|u - \Pi_h u\|_{\mathbf{L}^1(\Omega_h)} = \|u - \pi_{\mathcal{M}} u\|_{\mathbf{L}^1(\Omega)}$$

for all unit meshes $\mathcal{H}$ with respect to $(\mathcal{M}(\mathbf{x}))_{\mathbf{x}\in\Omega}$. The numerical examples in the end of this chapter will numerically demonstrate the efficiency of the continuous model. In particular, we will observe that

– the model is accurate, and the equivalence [4.21]≈[4.22] is observed for non-quadratic functions and non-constant continuous meshes;

– the error due to the fact that mesh generator generates edges with length not strictly equal to one is negligible. In particular, the range for the lengths of the edges given in definition 3.18 ensures reliable numerical results.

The constant 2 involved in theorem 4.2 arises in the perfect case where the mesh is only composed of perfect unit elements. Note this is only possible in 2D, whereas it is always false in 3D as it is impossible to tesselate the space by using the regular tetrahedron only. In a more practical situation, this constant needs to be evaluated from one unit mesh in order to estimate the deviation between the continuous mesh complexity with respect to the number of vertices of the unit mesh. In other words, the number of vertices N_v of a unit mesh satisfies

$$N_v = C\,N,$$

where N is the continuous mesh complexity and C a constant. The constant C depends on the domain shape, the mesh generator used and the unit mesh resulting quality. Consequently, it reflects how far the current mesh is from the perfect unity. Several examples are given in the numerical examples section and this relation is illustrated.

To conclude this section, the proof of theorem 4.2 is given. This proof is based on the exact expression of the continuous linear interpolate given by the following proposition:

PROPOSITION 4.2.– The continuous interpolate $\pi_{\mathcal{M}} u$ evaluated at $\mathbf{a} \in \Omega$ for a continuous mesh $(\mathcal{M}(\mathbf{x}))_{\mathbf{x}\in\Omega}$ and for a smooth function u is given by

$$\pi_{\mathcal{M}} u(\mathbf{a}) = p^*(0),$$

where p^* is the unique linear polynomial solution of

$$p^* = \min_{p \in \mathcal{P}^1} \|u_Q - p\|_{\mathbf{L}^2(\mathcal{B}_{\mathcal{M}})},$$

where u_Q is the quadratic model of u at $\mathbf{a}$ and $\mathcal{B}_{\mathcal{M}}$ is the unit ball of $\mathcal{M}$ at $\mathbf{a}$. $\pi_{\mathcal{M}}$ is given by

$$\pi_{\mathcal{M}} u(\mathbf{a}) = u(\mathbf{a}) + \nabla u(\mathbf{a}) + \frac{1}{c_d} \mathit{trace}(\mathcal{M}^{-\frac{1}{2}}(\mathbf{a})\, H(\mathbf{a})\, \mathcal{M}^{-\frac{1}{2}}(\mathbf{a})),$$

where c_d is a constant that depends only on the space dimension d:

$$c_2 = \frac{1}{8} \quad \text{and} \quad c_3 = \frac{1}{10}.$$

PROOF.– The quadratic model u_Q of u at point $\mathbf{a}$ defined by

$$u_Q(\mathbf{a}; \mathbf{x}) = u(\mathbf{a}) + \nabla u(\mathbf{a})(\mathbf{x} - \mathbf{a}) + \frac{1}{2}\langle(\mathbf{x} - \mathbf{a}), H(\mathbf{a})(\mathbf{x} - \mathbf{a})\rangle,$$

which becomes, after the translation $\mathbf{x} \mapsto \mathbf{x} + \mathbf{a}$,

$$u_Q(\mathbf{x}) = \frac{1}{2}\, {}^t\mathbf{x}\, H(\mathbf{a})\, \mathbf{x} + \nabla u(\mathbf{a})\, \mathbf{x} + u(\mathbf{a}).$$

The linear polynomial p^* is given by

$$p^* \ : \mathbf{x} \in \mathcal{B}_{\mathcal{M}} \mapsto {}^t\mathbf{g}\,\mathbf{x} + c,$$

where $c \in \mathbb{R}$ and $\mathbf{g} \in \mathbb{R}^3$. As the space of linear polynomials $\mathcal{P}^1$ and $\mathcal{B}_{\mathcal{M}}$ are convex, p^* exists and is unique. We seek for p^* verifying the following condition:

$$\forall p \in \mathcal{P}^1, \int_{\mathcal{B}_{\mathcal{M}}} (u_Q(\mathbf{x}) - p^*(\mathbf{x}))\, p(\mathbf{x})\, \mathbf{dx} \, = \, 0.$$

In particular, it is true for the following basis of $\mathcal{P}^1$: $\mathbf{x} \mapsto 1$, $\mathbf{x} \mapsto x_i$ for $i \in \{1,2,3\}$. The previous condition leads to

$$\begin{aligned} &\int_{\mathcal{B}_{\mathcal{M}}} (u_Q(\mathbf{x}) - p^*(\mathbf{x}))\, \mathbf{dx} = 0 \\ \text{and} \quad &\int_{\mathcal{B}_{\mathcal{M}}} (u_Q(\mathbf{x}) - p^*(\mathbf{x}))\, x_i\, \mathbf{dx} = 0 \quad \text{for } i = \{1,2,3\}. \end{aligned} \qquad [4.23]$$

The initial integration domain $\mathcal{B}_{\mathcal{M}}$ is mapped onto the reference continuous element unit ball $\mathcal{B}_{I_3}$ by using the following one-to-one change of variables:

$$\begin{aligned} \mathcal{B}_{\mathcal{M}} &\longrightarrow \mathcal{B}_{I_3}, \\ \mathbf{x} &\longmapsto \mathbf{y} = \mathcal{M}(\mathbf{a})^{\frac{1}{2}}\, \mathbf{x}. \end{aligned}$$

Functions u_Q and p^* become

$$\begin{aligned} u_Q(\mathbf{x}) &= \tilde{u}_Q(\mathbf{y}) = \tfrac{1}{2}\, {}^t\mathbf{y}\, \mathcal{M}(\mathbf{a})^{-\frac{1}{2}}\, H(\mathbf{a})\, \mathcal{M}(\mathbf{a})^{-\frac{1}{2}}\mathbf{y} + {}^t\mathbf{y}\, \mathcal{M}(\mathbf{a})^{-\frac{1}{2}}\, \nabla u(\mathbf{a}) + u(\mathbf{a}), \\ p^*(\mathbf{x}) &= \tilde{p}^*(\mathbf{y}) = {}^t\mathbf{y}\, \mathcal{M}(\mathbf{a})^{-\frac{1}{2}}\, \mathbf{g} + c. \end{aligned}$$

We now consider the following basis: $\mathbf{y} \mapsto 1$, $\mathbf{y} \mapsto y_i$, for $i \in \{1,2,3\}$. Equations [4.23] become

$$\begin{aligned} &\int_{\mathcal{B}_{I_3}} (\tilde{u}_Q(\mathbf{y}) - \tilde{p}^*(\mathbf{y}))\, \det(\mathcal{M}(\mathbf{a})^{-\frac{1}{2}})\, \mathbf{dy} = 0, \\ &\int_{\mathcal{B}_{I_3}} (\tilde{u}_Q(\mathbf{y}) - \tilde{p}^*(\mathbf{y}))\, y_i\, \det(\mathcal{M}(\mathbf{a})^{-\frac{1}{2}})\, \mathbf{dy} = 0 \end{aligned}$$

for $i = \{1, 2, 3\}$. A formal integration in 3D (Sykora 1974) gives

$$\left(\frac{2}{15}\,\mathrm{trace}(\mathcal{M}(\mathbf{a})^{-\frac{1}{2}}\,H(\mathbf{a})\,\mathcal{M}(\mathbf{a})^{-\frac{1}{2}}) + \frac{4}{3}\,(u(\mathbf{a}) - c)\right)\det(\mathcal{M}(\mathbf{a})^{-\frac{1}{2}}) = 0,$$

$$\left(\frac{4}{15}\,\mathcal{M}(\mathbf{a})^{-\frac{1}{2}}\,(\nabla u(\mathbf{a}) - \mathbf{g})\right)\det(\mathcal{M}(\mathbf{a})^{-\frac{1}{2}}) = 0,$$

from which the expression of p^* is deduced:

$$\mathbf{g} = \nabla u(\mathbf{a}) \text{ and } c = u(\mathbf{a}) + \frac{1}{10}\mathrm{trace}(\mathcal{M}(\mathbf{a})^{-\frac{1}{2}}\,H(\mathbf{a})\,\mathcal{M}(\mathbf{a})^{-\frac{1}{2}}).$$

Finally, for the two-dimensional case, the expression of p^* is given by

$$\mathbf{g} = \nabla u(\mathbf{a}) \text{ and } c = u(\mathbf{a}) + \frac{1}{8}\mathrm{trace}(\mathcal{M}(\mathbf{a})^{-\frac{1}{2}}\,H(\mathbf{a})\,\mathcal{M}(\mathbf{a})^{-\frac{1}{2}}). \; \square$$

The proof of theorem 4.2 is deduced from the definition of $\pi_{\mathcal{M}} u(\mathbf{a})$ in propostion 4.2. □

In the case where the continuous mesh is constant and the function u quadratic, we verify [4.21] = [4.22]. Notice that using the $\mathbf{L}^2$-projection of the quadratic model u_Q of u is necessary to ensure the specific equivalence between the discrete linear interpolate Π_h and the continuous linear interpolate $\pi_{\mathcal{M}}$ of theorem 4.2. However, the continuous linear interpolate is still well defined if we use the function u instead of u_Q. It seems then possible to define the continuous linear interpolate for far less regular functions. For instance, one may consider functions that are only locally $\mathbf{L}^2$. An open problem is then to find a discrete linear interpolation operator, which enables discrete and discontinuous approaches to be linked. Some works involving other interpolation operators have already been considered; we can cite the developments in Formaggia and Perotto (2001) that derive an interpolation error estimate based on the Clément's interpolate (Clément 1975).

4.3. Numerical validation of the continuous interpolation error

Using the continuous mesh framework, the continuous linear interpolation error in the $\mathbf{L}^1$ norm can be analytically computed for any analytic function u given on any continuous mesh $\mathbf{M} = (\mathcal{M}(\mathbf{x}))_{\mathbf{x}\in\Omega}$ of a domain Ω. This calculus does not require any discrete support, for example, any mesh. In this section, this is emphasized on several examples in two dimensions (2D) and three dimensions (3D). To validate the approach, each calculated continuous interpolation error is compared to the discrete interpolation error computed on a unit mesh with respect to the continuous one.

4.3.1. *Continuous interpolation error calculation*

Embedded continuous meshes. We consider the set of continuous embedded meshes $\mathbf{M}_1(\alpha) = (\mathcal{M}_{1,\alpha}(\mathbf{x}))_{\mathbf{x}\in\Omega_1}$ defined on square domain $\Omega_1 = [0,1] \times [0,1]$ and given by

$$\mathcal{M}_{1,\alpha}(x,y) = \alpha \begin{bmatrix} h_1^{-2}(x,y) & 0 \\ 0 & h_2^{-2}(x,y) \end{bmatrix}, \qquad [4.24]$$

where $h_1(x,y) = 0.1(x+1) + 0.05(x-1)$ and $h_2(x,y) = 0.2$. The parameter α is used to control the level of accuracy of the mesh. The continuous mesh $\mathbf{M}_1(\alpha)$ becomes coarser when α decreases, but anisotropic quotients and orientations remain constant. This trend is given by the computation of the complexity $\mathcal{C}(\mathbf{M}_1(\alpha))$:

$$\mathcal{C}(\mathbf{M}_1(\alpha)) = N(\alpha) = \iint_{\Omega_1} \frac{1}{h_1 h_2}(x,y)\,\mathrm{d}x\mathrm{d}y = \frac{200}{3}\ln(2)\alpha.$$

The parameter α (or N) defines embedded continuous meshes according to definition 3.14.

The continuous interpolation error on $\mathbf{M}_1(\alpha)$ is now computed for two analytical functions u_1 and u_2:

- u_1 is a quadratic function given by $u_1(x,y) = 6x^2 + 2xy + 4y^2$;
- u_2 is a non-quadratic function given by $u_2(x,y) = \mathrm{e}^{(2x^2+y)}$.

With regard to function u_1, the pointwise continuous interpolation error on $\mathbf{M}_1(\alpha)$ is

$$\begin{aligned}(u_1 - \pi_{\mathcal{M}_{1,\alpha}} u_1)(x,y) &= \frac{1}{8}\,\mathrm{trace}\left(\mathcal{M}_{1,\alpha}^{-\frac{1}{2}}(x,y)\,|H_{u_1}(x,y)|\,\mathcal{M}_{1,\alpha}^{-\frac{1}{2}}(x,y)\right) \\ &= \frac{27\,x^2 + 18\,x + 35}{800\,\alpha}.\end{aligned}$$

The previous expression is then integrated over Ω_1:

$$\iint_{\Omega_1} |u_1 - \pi_{\mathcal{M}_{1,\alpha}} u_1|(x,y)\,\mathrm{d}x\mathrm{d}y = \frac{53}{800\,\alpha} = \frac{53\,\ln(2)}{12\,N(\alpha)}.$$

For function u_2, the pointwise continuous interpolation error on $\mathbf{M}(\alpha)$ is

$$(u_2 - \pi_{\mathcal{M}_{1,\alpha}} u_2)(x,y) = \frac{1}{8} \operatorname{trace}\left(\mathcal{M}_{1,\alpha}^{-\frac{1}{2}}(x,y) \, |H_{u_2}(x,y)| \, \mathcal{M}_{1,\alpha}^{-\frac{1}{2}}(x,y) \right)$$
$$= \frac{e^{4x^2+y}}{8\,\alpha} \left((0.15x + 0.05)^2 \, (4 + 16x^2) + 0.05 \right).$$

By a direct integration over Ω_1, we get

$$\iint_{\Omega_1} |u_2 - \pi_{\mathcal{M}_{1,\alpha}} u_2|(x,y) \, \mathrm{d}x\mathrm{d}y \approx \frac{0.2050950191}{\alpha} \approx \frac{13.673 \ln(2)}{N(\alpha)}.$$

Some remarks can be made before comparing the continuous evaluation of interpolation error to the discrete one, which will be done in the next section.

Convergence of embedded meshes. According to the previous examples, if given a continuous mesh $\mathbf{M}(\alpha)$ and a smooth function u, the continuous interpolation for u on $\mathbf{M}(\alpha)$ is written in a generic way:

$$\|u - \pi_{\mathcal{M}} u\|_{\mathbf{L}^1(\Omega_1)} = \frac{C_u}{N(\alpha)},$$

where C_u depends on u and $N(\alpha)$ is the complexity of $\mathbf{M}(\alpha)$. The previous expression gives quantitative information on the order of convergence of the interpolation error on a sequence of continuous embedded meshes issued from $\mathbf{M}(1)$. Indeed, a simple analogy with uniform meshes leads to put $N(\alpha) = O(h^{-2}(\alpha))$, so that

$$\|u - \pi_{\mathcal{M}} u\|_{\mathbf{L}^1(\Omega_1)} = C'_u \, h^2(\alpha).$$

Consequently, the continuous interpolation error model predicts an order of convergence of two on a sequence of embedded continuous meshes. From the discrete view, it is well known that a uniform refinement leads to a second order of convergence for the linear interpolation error with respect to a smooth function. This fact is also given by the continuous analysis. With this framework, the prediction of the rate of convergence is generalized to anisotropic meshes, because of the use of the continuous complexity. Indeed, a unique real parameter h as used classically (Binev et al. 2004) is no more sufficient to take into account anisotropic features for a sequence of meshes.

Non-embedded continuous meshes. Now let us give a more complex example in order to demonstrate the ability of the continuous mesh model to predict the order of convergence. In this example, a discrete study of the prediction of the interpolation error is impossible, whereas a clear convergence order is exhibited with the continuous mesh model. We consider the following set of continuous meshes $\mathbf{M}_2(\alpha) = (\mathcal{M}_{2,\alpha}(\mathbf{x}))_{\mathbf{x}\in\Omega_2}$ defined on square domain $\Omega_2 = [0.5, 1] \times [0.5, 1]$ and given by

$$\mathcal{M}_{2,\alpha}(x, y) = F(x, y) \begin{bmatrix} \alpha^2 h_1^{-2}(x, y) & 0 \\ 0 & \alpha h_2^{-2}(x, y) \end{bmatrix} {}^tF(x, y),$$

where

$$F(x, y) = \frac{1}{\sqrt{x^2 + y^2}} \begin{bmatrix} x & -y \\ y & x \end{bmatrix},$$

$$h_1^{-2}(x, y) = 4(x^2 + y^2) \quad \text{and} \quad h_2(x, y)^{-2} = \frac{1}{2\sqrt{x^2 + y^2}}.$$

Note that this set is no more embedded according to definition 3.14. On the contrary, the continuous meshes are rather nonuniformly embedded, as one size is scaled by $1/\alpha$ and the other one is scaled by $1/\sqrt{\alpha}$. The equivalent discrete refinement process is no more homogeneous, that is, the factor of division of each edge while increasing α depends on the edge coordinates. Consequently, the order of convergence seems *a priori* unpredictable contrary to the embedded case. However, we show that we are able to predict the asymptotic convergence order of the continuous interpolation error for the set spanned by $\mathbf{M}_2(\alpha)$ by using the continuous analysis. The complexity of $\mathbf{M}_2(\alpha)$ is given by

$$\mathcal{C}(\mathbf{M}_2(\alpha)) = N(\alpha) = \iint_{\Omega_2} \frac{1}{h_1 h_2}(x, y)\, \mathrm{d}x\mathrm{d}y \approx 0.364\, \alpha^{\frac{3}{2}}.$$

We consider the interpolation error of the quadratic function $u_3(x, y) = x^2 + y^2$. The pointwise continuous interpolation error on $\mathbf{M}_2(\alpha)$ is given by

$$\begin{aligned}(u_3 - \pi_{\mathcal{M}_{2,\alpha}} u_3)(x, y) &= \frac{1}{8} \operatorname{trace}\left(\mathcal{M}_{2,\alpha}^{-\frac{1}{2}}(x, y)\, |H_{u_3}(x, y)|\, \mathcal{M}_{2,\alpha}^{-\frac{1}{2}}(x, y)\right) \\ &= \frac{1}{16\,(x^2 + y^2)\,\alpha^2} + \frac{\sqrt{x^2 + y^2}}{2\,\alpha}.\end{aligned}$$

The previous expression is then integrated over Ω_2; it results in

$$\iint_{\Omega_2} |u_3 - \pi_{\mathcal{M}_{2,\alpha}} u_3|(x,y)\,\mathrm{d}x\mathrm{d}y = \frac{0.133}{N(\alpha)^{\frac{4}{3}}} + \frac{0.014}{N(\alpha)^{\frac{2}{3}}}.$$

The inhomogeneity in the scaling of the sizes leads to two terms with different orders of convergence: $\frac{8}{3}$ and $\frac{4}{3}$ for the first and the second terms, respectively. Consequently, the asymptotic order of convergence for the continuous interpolation error of u_3 on $\mathbf{M}_2(\alpha)$ is only $O(N^{-\frac{2}{3}})$, leading to a convergence order of $\frac{4}{3} \approx 1.33$. This is less than the second order reached on the previous set of meshes defined by $\mathbf{M}_1(\alpha)$. Indeed, for a sufficiently large value of N, the term of order $O(N^{-\frac{4}{3}})$ becomes negligible with respect to the low order term $O(N^{-\frac{2}{3}})$. However, this approximation is only true asymptotically. Practically, the complexity allowing this simplification depends on the constant 0.014 and 0.133. According to Figure 4.1 (left), as soon as the complexity becomes greater than 1000, the asymptotic order of convergence is fully represented by $\frac{0.014}{N(\alpha)^{\frac{2}{3}}}$. Note that this value is reachable in practice on discrete meshes.

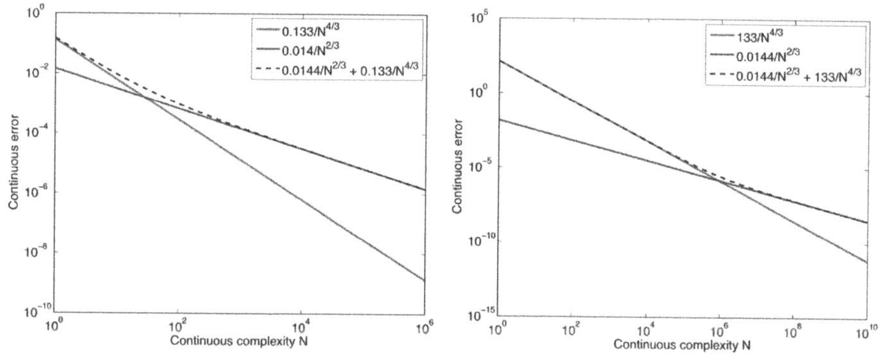

Figure 4.1. *Left: Convergence history obtained for the function u_3 on the set of inhomogeneous continuous meshes spans by $\mathrm{M}_2(\alpha)$ defined on Ω_2. Right: Convergence history for the error function of remark 4.5. In that specific case, the coefficients involved in the error expression lead to observe $O(N^{-\frac{4}{3}})$ as a predominant term in the reachable range of complexity $[1, 10^6]$, whereas the asymptotic convergence order is $O(N^{-\frac{2}{3}})$. For a color version of this figure, see www.iste.co.uk/dervieux/meshadaptation1*

REMARK 4.5.– Depending on the constants involved in the estimation of the order of convergence, a convergence order different from the asymptotic one can be observed. If we suppose that the error for a function u on $\mathbf{M}_2(\alpha)$ is written as

$$\iint_{\Omega_2} |u - \pi_{\mathcal{M}_{2,\alpha}} u|(x,y)\,\mathrm{d}x\mathrm{d}y = \frac{133}{N(\alpha)^{\frac{4}{3}}} + \frac{0.014}{N(\alpha)^{\frac{2}{3}}},$$

then the asymptotic complexity such that the term of order $\frac{4}{3}$ becomes negligible is difficult to reach in practice with discrete meshes (greater than 10^6; see Figure 4.1, right). Consequently, the observed order of convergence for acceptable complexities ($\mathcal{C}(\mathbf{M}_2(\alpha)) \in [1, 10^6]$) is of the order of $O(N^{-\frac{4}{3}})$. This example shows that predicting both the order of convergence and the magnitude of the constants is crucial to get a reliable asymptotic prediction of the interpolation error. □

In the next section, the analytic evaluation of the constant C_u along with the convergence order are compared to discrete estimations obtained by generating unit discrete meshes with respect to $\mathbf{M}_1(\alpha)$ and $\mathbf{M}_2(\alpha)$ for different values of α.

4.3.2. Comparison with discrete interpolation error computation

Unit meshes. To validate the previous continuous evaluation of the interpolation error, a series of discrete unit meshes with respect to $\mathbf{M}_1(\alpha) = (\mathcal{M}_{1,\alpha}(\mathbf{x}))_{\mathbf{x}\in\Omega_1}$ is generated. These meshes are considered for $\alpha = \{1, 2, 4, 8, 16, 32\}$. We denote by $\{\mathcal{H}_{1,\alpha}\}_{\alpha\in[1...32]}$ this sequence of discrete meshes. They have been generated using `Yams` (Frey 2001). Figure 4.2 depicts some of these meshes. Histograms reporting for each mesh the length of its edges and the quality of its elements are given in Table 4.1. These histograms point out the gap between the generated unit meshes and a perfect unit mesh. We notice that an almost perfect quality is reached for each mesh and that more than 80% of the length of edges lie in the range $[\frac{\sqrt{2}}{2}, \sqrt{2}]$ as soon as $\alpha \geq 4$. Unit meshes sequence $\{\mathcal{H}_{2,\alpha}\}_{\alpha\in[16...512]}$ with respect to $\mathbf{M}_2(\alpha)$ are depicted in Figure 4.3. Similar conclusions arise.

Complexity versus number of vertices. We first study the correlation between the discrete number of vertices N_v and the continuous complexity N. This preliminary study is necessary in the continuous discrete comparison. Indeed, the continuous interpolation error estimate involves the continuous complexity, whereas the discrete number of vertices is used in the discrete error computation. With regard to the sequence of discrete meshes $\{\mathcal{H}_{1,\alpha}\}_\alpha$ with respect to $\{\mathbf{M}_1(\alpha)\}_\alpha$, the discrete number of nodes N_v is plotted as a function of the complexity N in Figure 4.4(top right). This function is linear, and the slope evaluated numerically gives

$$\forall \mathcal{H}_\alpha, \quad N_v(\alpha) = 1.54\,N(\alpha).$$

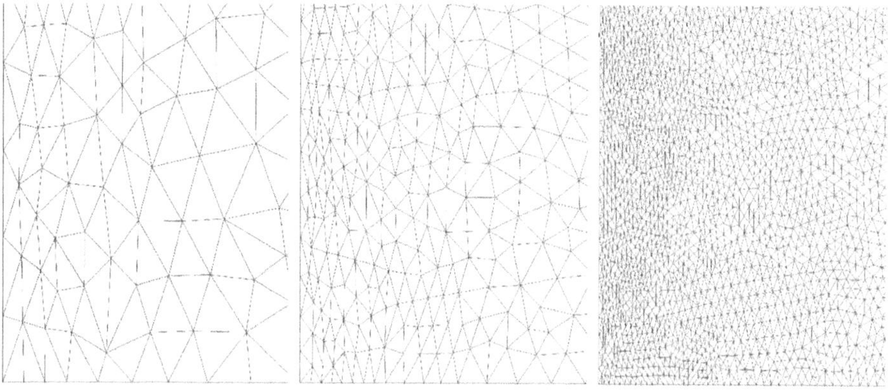

Figure 4.2. *Unit meshes $\{\mathcal{H}_{1,\alpha}\}_\alpha$ with respect to $\{\mathbf{M}_1(\alpha)\}_\alpha$ on Ω_1 for $\alpha = \{1, 8, 32\}$ from left to right*

	$\mathcal{H}_{1,1}$		$\mathcal{H}_{1,2}$		$\mathcal{H}_{1,4}$		$\mathcal{H}_{1,8}$		$\mathcal{H}_{1,16}$		$\mathcal{H}_{1,32}$	
$0.20 < L < 0.50$	5	1.94 %	1	0.19 %	6	0.71 %			9	0.28 %	8	0.12 %
$0.50 < L < 0.71$	53	20.54 %	130	25.15 %	82	9.69 %	281	16.6 %	276	8.44 %	905	13.69 %
$0.71 < L < 0.90$	116	44.96 %	204	39.46 %	352	41.61 %	564	33.31 %	1436	43.89 %	2491	37.67 %
$0.90 < L < 1.11$	68	26.36 %	133	25.73 %	279	32.98 %	542	32.01 %	1071	32.73 %	2081	31.47 %
$1.11 < L < 1.41$	16	6.20 %	49	9.48 %	127	15.01 %	306	18.07 %	480	14.67 %	1127	17.04 %
$1 < Q < 2$	160	99.38 %	328	99.39 %	541	99.27 %	328	99.18 %	328	99.77 %	328	99.70 %
$2 < Q < 3$	1	0.62 %	2	0.61 %	4	0.73 %	2	0.82 %	2	0.23 %	2	0.30 %

Table 4.1. *Quality $1/Q_{\mathcal{M}_1}$ and length L of the edges for the unit meshes $\{\mathcal{H}_{1,\alpha}\}_\alpha$ with respect to $\{\mathbf{M}_1(\alpha)\}_\alpha$ in 2D for $\alpha = \{1, 2, 4, 6, 8, 16, 32\}$, cf. (4.24)*

Note that this constant handles the discrepancy between the perfect unit mesh and the generated constrained discrete mesh with respect to the continuous one. Constraints arise from the domain boundary, the mesh generator which is used, and from the smoothness of the continuous mesh. This constant is in the theoretical framework equal to 2 (see Theorem 3.3 in Loseille et al. (2010a)). When the constant is equal to 2, it ensures the following equality between discrete and continuous errors:

$$\frac{C_u}{N} = 2\frac{C_u}{N_v}.$$

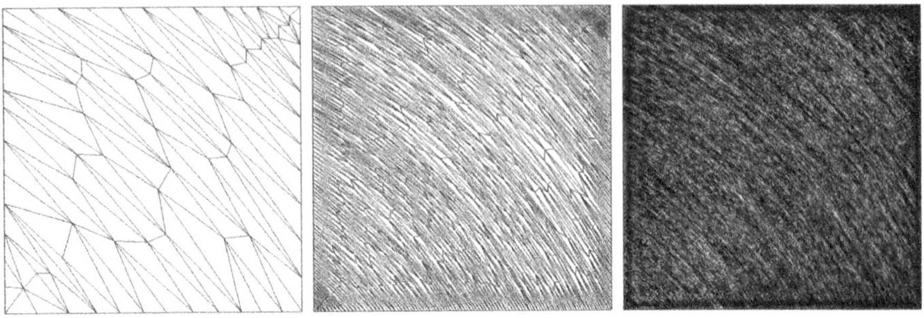

Figure 4.3. *Unit meshes $\{\mathcal{H}_{2,\alpha}\}_\alpha$ with respect to $\{\mathbf{M}_2(\alpha)\}_\alpha$ on Ω_2 for $\alpha = \{16, 128, 512\}$ from left to right*

In the case of the set of inhomogeneous continuous meshes defined by $\mathbf{M}_2(\alpha)$, the constant is equal to 3.37.

Discrete interpolation error. The discrete interpolation error of a function u on a mesh $\mathcal{H}_\alpha$ reads

$$\|u - \Pi_h u\|_{\mathbf{L}^1(\mathcal{H}_\alpha)} = \sum_{K \in \mathcal{H}_\alpha} \|u - \Pi_h u\|_{\mathbf{L}^1(K)},$$

where K stands for an element of $\mathcal{H}_\alpha$. In our experiments, the elementwise interpolation error $\|u - \Pi_h u\|_{\mathbf{L}^1(K)}$ is computed by means of a fifth-order Gauss quadrature numerical integration. The discrete interpolation error is compared to the continuous one by considering N_v instead of N in the continuous estimates. Figure 4.4 plots discrete and continuous interpolation errors for functions u_1 and u_2 on $\mathbf{M}_1(\alpha)$ and for function u_3 on $\mathbf{M}_2(\alpha)$. An excellent correlation is obtained between the continuous and the discrete evaluations. The slight differences between the continuous and the discrete expressions come from the practical difficulty of generating a perfect unit mesh. Indeed, mesh generators generate unit meshes in the sense of definition 3.15. It implies that elements are quasi-unit. In consequence, most of the length of edges lie in the range $[\frac{\sqrt{2}}{2}, \sqrt{2}]$. Bounds of the interpolation error can be deduced from this tolerance. For every function u and for every continuous mesh, we have

$$\|u - \pi_{\mathcal{M}_{\min}} u\|_{\mathbf{L}^1} \le \|u - \pi_{\mathcal{M}} u\|_{\mathbf{L}^1} \le \|u - \pi_{\mathcal{M}_{\max}} u\|_{\mathbf{L}^1},$$

where $\mathcal{M}_{\min} = 2\,\mathcal{M}$ and $\mathcal{M}_{\max} = \frac{1}{2}\,\mathcal{M}$. Unit meshes composed of edges with a length lying in the acceptable range will verify

$$\frac{1}{2}\,\|u - \pi_{\mathcal{M}} u\|_{\mathbf{L}^1} \leq \|u - \Pi_h u\|_{\mathbf{L}^1} \leq 2\,\|u - \pi_{\mathcal{M}} u\|_{\mathbf{L}^1}.$$

These bounds are plotted in Figure 4.4. Note that previous bounds impact only the constant evaluation C_u and not the order of convergence of the interpolation error.

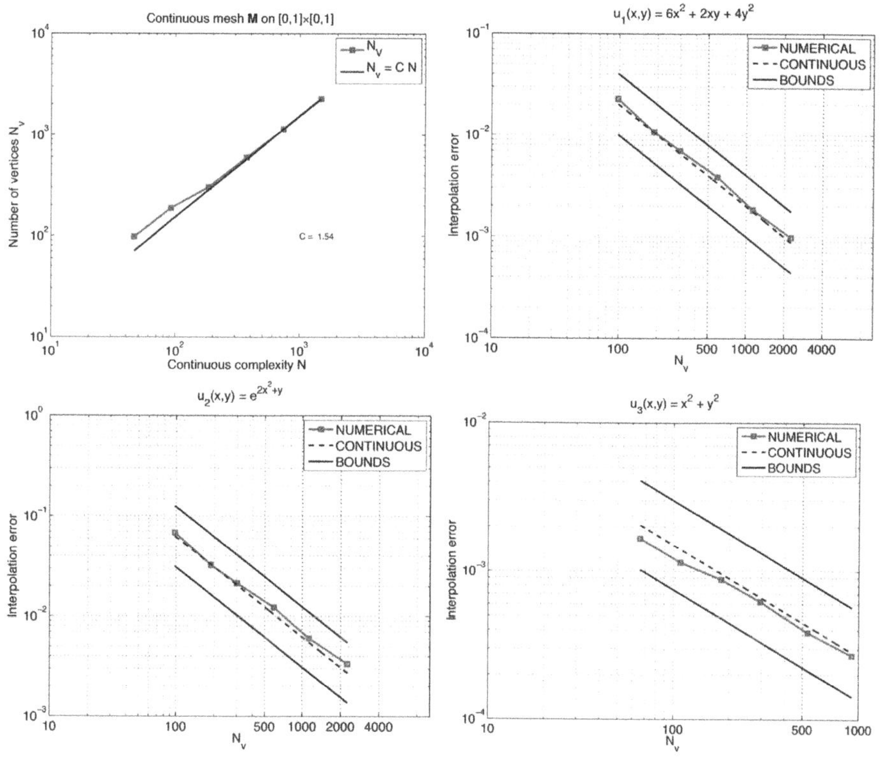

Figure 4.4. *Top left, comparison between the continuous mesh complexity and the number of vertices for a sequence of unit meshes $\{\mathcal{H}_{1,\alpha}\}_\alpha$ with respect to $\{\mathbf{M}_1(\alpha)\}_\alpha$. The remaining panels show a comparison between continuous interpolation error $\|u - \pi_{\mathcal{M}} u\|_{\mathbf{L}^1(\Omega)}$ and discrete interpolation error $\|u - \Pi_h u\|_{\mathbf{L}^1(\Omega_h)}$ for function u_1 (top right), u_2 (bottom left) on $\mathbf{M}_1(\alpha)$ and u_3 (bottom right) on $\mathbf{M}_2(\alpha)$. Black plain lines represent continuous interpolation errors when considering $\mathcal{M}_{\max}$ and $\mathcal{M}_{\min}$. For a color version of this figure, see www.iste.co.uk/dervieux/meshadaptation1*

4.3.3. *Three-dimensional validation*

An analytical example. We first consider the set of three-dimensional embedded continuous meshes $\mathbf{M}_3(\alpha) = (\mathcal{M}_{3,\alpha}(\mathbf{x}))_{\mathbf{x}\in\Omega_3}$ defined on the domain $\Omega_3 = [0,1] \times [0,1] \times [0,1]$, which are given by

$$\mathcal{M}_{3,\alpha}(x,y,z) = \alpha \begin{bmatrix} h_1^{-2}(x,y,z) & 0 & 0 \\ 0 & h_2^{-2}(x,y,z) & 0 \\ 0 & 0 & h_3^{-2}(x,y,z) \end{bmatrix},$$

where $h_1(x,y,z) = 0.1(x+1) + 0.05(x-1)$, $h_2(x,y,z) = 0.2$, and $h_3(x,y,z) = 0.2(z+2)$. The parameter α is used to control the level of accuracy of the mesh. The continuous mesh becomes coarser when α decreases. This trend is given by the computation of the complexity $\mathcal{C}(\mathbf{M}_3(\alpha))$:

$$\begin{aligned} \mathcal{C}(\mathbf{M}_3(\alpha)) &= N(\alpha) = \iiint_{\Omega_3} \frac{1}{h_1 h_2 h_3}(x,y,z)\, \mathrm{d}x\mathrm{d}y\mathrm{d}z \\ &= \frac{1000}{3} \ln(2)\big(\ln(3) - \ln(2)\big)\alpha^{\frac{3}{2}}. \end{aligned}$$

We consider the interpolation error of the function $u_4(x,y,z) = e^{2x+y+z}$. The pointwise continuous linear interpolation error is

$$\begin{aligned} &(u_4 - \pi_{\mathcal{M}_{3,\alpha}} u_4)(x,y,z) \\ &= \frac{1}{10^5} \frac{e^{2\,x+y+z}\left(441\,x^2 + 798\,x + 2361 + 400\,z^2 + 1600\,z\right)}{\alpha}. \end{aligned}$$

Integration over Ω_3 leads to

$$\begin{aligned} &\iiint_{\Omega_3} |u_4 - \pi_{\mathcal{M}_{3,\alpha}} u_4|(x,y,z)\, \mathrm{d}x\mathrm{d}y\mathrm{d}z \\ &= \frac{\left(-2133 - 1772\,e^4 + 7466\,e^2 - 10322\,e^6 + 6761\,e^8\right)}{4.\,10^5\,e^4\,\alpha} \\ &\approx \frac{0.73}{\alpha} \approx \frac{126.215}{N(\alpha)^{\frac{2}{3}}}. \end{aligned}$$

A sequence of unit meshes with respect to $\mathbf{M}_3(\alpha)$ for $\alpha \in \{2, 4, 8, 16, 32\}$ have been generated using `Yams` (Frey 2001) and `Gamanic3D` (George 2001) (see Figure 4.5). In this example, the constant linking the number of vertices and the continuous complexity is 2.07 (see Figure 4.6, left). Consequently, the perfect theoretical case is almost reached. The comparison between discrete and continuous interpolation error is depicted in Figure 4.6 (right). As for the two-dimensional examples, we observe an excellent correlation between the continuous model and the numerical computation.

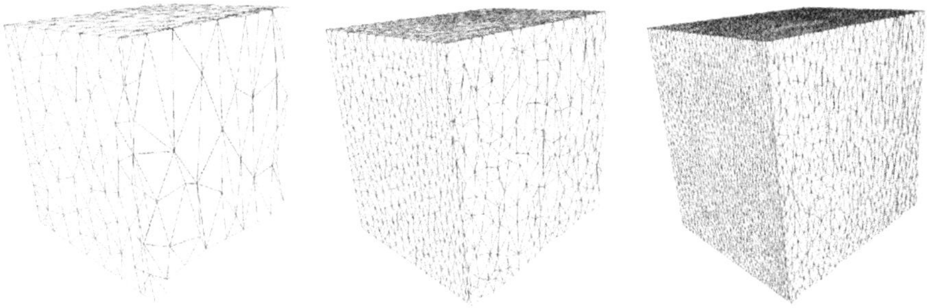

Figure 4.5. *Three-dimensional unit meshes $\{\mathcal{H}_{3,\alpha}\}_\alpha$ with respect to $\{\mathrm{M}_3(\alpha)\}_\alpha$ for $\alpha = \{1, 8, 32\}$ from left to right*

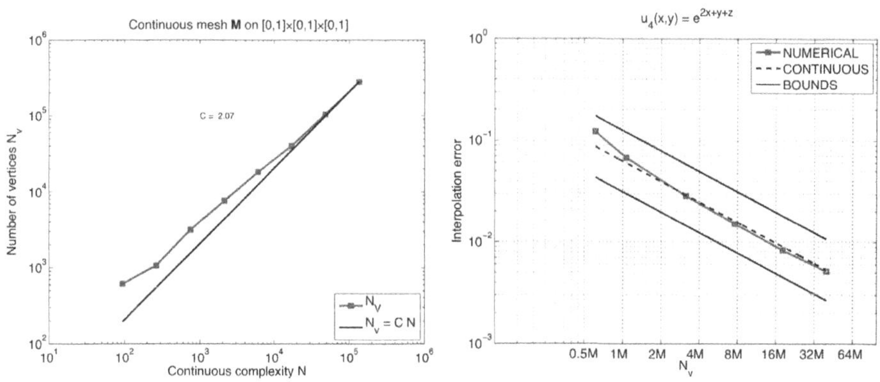

Figure 4.6. *Left: Comparison between the continuous mesh complexity and the number of vertices for a sequence of unit meshes $\{\mathcal{H}_{3,\alpha}\}_\alpha$ with respect to $\{\mathrm{M}_3(\alpha)\}_\alpha$ on Ω_3. Right: Comparison between the continuous interpolation error $\|u_4 - \pi_{\mathcal{M}_{3,\alpha}} u_4\|_{\mathbf{L}^1(\Omega_3)}$ and the discrete interpolation error $\|u_4 - \Pi_h u_4\|_{\mathbf{L}^1(\Omega_{3,h})}$. For a color version of this figure, see www.iste.co.uk/dervieux/meshadaptation1*

A numerical example. The theory is now applied to a more realistic three-dimensional example where the continuous interpolation error results from numerical computations and not from an analytical evaluation, which is not possible in that case. More precisely, the function and the continuous mesh (i.e. the metric field) are linearly interpolated on a discrete mesh. Then the continuous interpolation error is evaluated using these discrete data. This case is more representative of the practical use of the continuous mesh model. The aim of this example is to point out that, even with numerical computations, the continuous mesh model turns out to be a reliable model to predict interpolation error. We first define the continuous mesh $\mathbf{M}_4(\alpha) = (\mathcal{M}_{4,\alpha}(\mathbf{x}))$ on the domain $\Omega_4 = [-1,1] \times [-1,1] \times [-1,1] \subset \mathbb{R}^3$. The analytical expression of $\mathcal{M}_{4,\alpha}$ is inherited from the following quadratic form:

$$q(x,y,z) = 2x^2 - yx + 6y^2 + z^2.$$

The gradient vector is denoted $\mathbf{g}(x,y,z) = {}^t(\frac{\partial q}{\partial x}, \frac{\partial q}{\partial y}, \frac{\partial q}{\partial z})$. The local Frenet frame is given by $F = (\mathbf{n}, \mathbf{t}_1, \mathbf{t}_2)$, where the vectors $\mathbf{t}_1, \mathbf{t}_2$ lie in the plane orthogonal to $\mathbf{n} = \mathbf{g}/\|\mathbf{g}\|_2$. We assume that $(\mathbf{t}_1, \mathbf{t}_2)$ is an orthonormal basis of $\mathbb{R}^2$. We also consider the Hessian matrix H_q of q:

$$H_q = \begin{bmatrix} 4 & -1 & 0 \\ -1 & 12 & 0 \\ 0 & 0 & 2 \end{bmatrix}.$$

From this matrix, we define the submatrix

$$\overline{H}_q = \begin{bmatrix} {}^t\mathbf{t}_1\, H_q\, \mathbf{t}_1 & {}^t\mathbf{t}_1\, H_q\, \mathbf{t}_2 \\ {}^t\mathbf{t}_1\, H_q\, \mathbf{t}_2 & {}^t\mathbf{t}_2\, H_q\, \mathbf{t}_2 \end{bmatrix},$$

which is the projection of H_q onto the plane defined by $(\mathbf{t}_1, \mathbf{t}_2)$. We denote by $\overline{\lambda_1}$ and $\overline{\lambda_2}$ the eigenvalues of $\overline{H}_q$. The continuous mesh $\mathbf{M}_4(\alpha)$ parameterized by α is then given by

$$\mathcal{M}_{4,\alpha} = F \begin{bmatrix} \alpha^2\, \|\mathbf{g}\|_2^2 & & \\ & \alpha\, \dfrac{|\overline{\lambda_1}|}{2\,\|\mathbf{g}\|_2} & \\ & & \alpha\, \dfrac{|\overline{\lambda_2}|}{2\,\|\mathbf{g}\|_2} \end{bmatrix} {}^tF.$$

The complexity of $\mathbf{M}_4(\alpha)$ is

$$\mathcal{C}(\mathbf{M}_4(\alpha)) = \frac{\alpha^2}{2} \int_\Omega \sqrt{\|\mathbf{g}\|_2^2\, |\overline{\lambda_1}|\, |\overline{\lambda_2}|},$$

which can be evaluated numerically. This continuous mesh has the physical feature to follow the iso-surfaces of the function q and to adapt the sizes to the local curvature of the iso-surfaces. The iso-surfaces of q are represented in Figure 4.7(left). A sequence of unit meshes $\{\mathcal{H}_{4,\alpha}\}_\alpha$ with respect to $\{\mathbf{M}_4(\alpha)\}_\alpha$ for a complexity of

[4,000, ,8,000, 16,000, 32,000, 64,000, 128,000]

has been generated. Resulting unit meshes are depicted in Figure 4.8. In this case, the number of vertices is almost 2.3 times the continuous complexity.

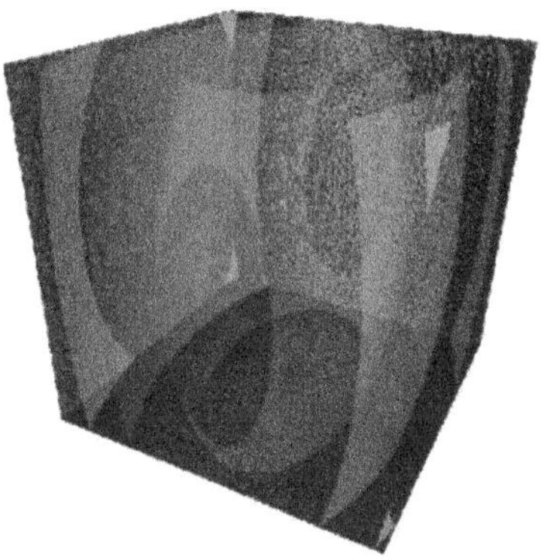

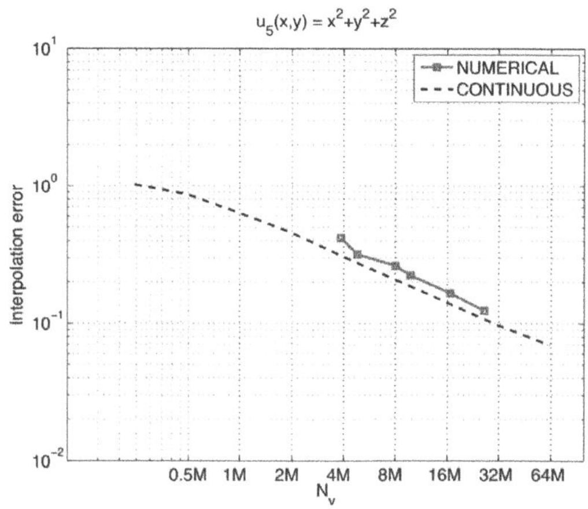

Figure 4.7. *Left: Iso-surfaces of function q. Right: Comparison between the continuous interpolation error $\|u_5 - \pi_{\mathcal{M}_{4,\alpha}} u_5\|_{\mathbf{L}^1(\Omega_4)}$ and the discrete interpolation error $\|u_5 - \Pi_h u_5\|_{\mathbf{L}^1(\Omega_{4,h})}$. In this case, the continuous interpolation error is computed numerically by means of the linear approximation of the continuous mesh on a fine uniform discrete mesh. For a color version of this figure, see www.iste.co.uk/dervieux/meshadaptation1*

We consider the interpolation error of the quadratic function

$$u_5(x, y, z) = x^2 + y^2 + z^2.$$

In this case, the continuous interpolation error on the continuous mesh $\mathbf{M}_4(\alpha) = (\mathcal{M}_{4,\alpha}(\mathbf{x}))_{\mathbf{x}\in\Omega}$ cannot be calculated analytically anymore. We propose to compute it numerically on a given discrete mesh $\mathcal{H}$ as follows:

$$\int_\Omega |u_5(\mathbf{x}) - \pi_{\mathcal{M}_{4,\alpha}} u_5(\mathbf{x})| \, \mathrm{d}\mathbf{x}$$
$$\approx \frac{1}{10} \sum_{K\in\mathcal{H}} |K| \,\mathrm{trace}\left(\mathcal{M}_{4,\alpha}(K)^{-\frac{1}{2}} |H_{u_5}(K)| \mathcal{M}_{4,\alpha}(K)^{-\frac{1}{2}}\right), \qquad [4.25]$$

where $\mathcal{M}_{4,\alpha}(K)$ and $|H_{u_5}(K)|$ are the linear discrete representation on the element K of the continuous mesh and the Hessian of u_5; thus, a simple first-order Gauss quadrature numerical integration is considered. The discrete interpolation error of function u_5 is computed on the sequence of discrete meshes $\{\mathcal{H}_{4,\alpha}\}_\alpha$. These discrete errors are compared to continuous interpolation errors evaluated on a fine uniform mesh of constant size $h = 0.004$ using relation [4.25]. This size corresponds to the minimal size prescribed by the continuous mesh $\mathbf{M}_4$(128,000). The correlation between the continuous interpolation error and the discrete interpolation error is shown in Figure 4.7 (right). As expected, the order of convergence is well predicted, but the numerical computations induce an error overestimation. Consequently, if the slope (i.e. the convergence order) of the error is well approximated, the perfect fitting due the approximated evaluations of C_u results in a slightly shifted curve.

Figure 4.8. *From left to right, views (cut through the plane $y = 0$) of the unit meshes $\{\mathcal{H}_{4,\alpha}\}_\alpha$ with respect to $\{\mathbf{M}_4(\alpha)\}_\alpha$ for a complexity of* 4,000, 8,000 and 16,000. *For a color version of this figure, see www.iste.co.uk/dervieux/meshadaptation1*

4.3.4. *Some conclusions*

These examples reveal that the interpolation error can be computed continuously without any discrete support. From this point of view, a discrete mesh is simply the projection of the continuous one. Discrepancies between continuous and discrete interpolation errors are due to projection errors. Practically, this gap depends on the mesh generator used and on the difficulty to generate the desired unit mesh. When the continuous interpolation error cannot be evaluated analytically anymore, we have shown that a numerical evaluation using a discrete mesh still provides accurate predictions of either the convergence order or the error magnitude. Using this accurate and generic continuous interpolation error, we address in the next section the problem of minimizing the interpolation error in $\mathbf{L}^p$ norm on Ω.

4.4. Optimal control of the interpolation error in $\mathbf{L}^p$ norm

Using the definition of the linear continuous interpolate $\pi_{\mathcal{M}}$ given by corollary 4.1, the following 3D pointwise interpolation error for u on $\mathbf{M} = (\mathcal{M}(\mathbf{x}))_{\mathbf{x}\in\Omega}$ is obtained:

$$e_{\mathcal{M}}(\mathbf{x}) = (u - \pi_{\mathcal{M}} u)(\mathbf{x}) = \frac{1}{10}\sum_{i=1}^{3} h_i^2(\mathbf{x})\, {}^t\mathbf{v}_i(\mathbf{x})\, |H_u(\mathbf{x})|\, \mathbf{v}_i(\mathbf{x}),$$

where H_u is the Hessian of u, $(\mathbf{v}_i)_{i=1,3}$ are the local eigen-directions of $\mathbf{M}$, and $(h_i)_{i=1,3}$ are the local sizes of $\mathbf{M}$ along these directions. It is then possible to set the well-posed global optimization problem of finding the optimal continuous mesh minimizing the continuous interpolation error in $\mathbf{L}^p$ norm as follows:

$$\text{Find } \mathbf{M}_{\mathbf{L}^p} = \min_{\mathbf{M}} E_p(\mathbf{M}) = \min_{\mathbf{M}} \left(\int_\Omega e_{\mathcal{M}}^p \right)^{\frac{1}{p}} = \min_{\mathbf{M}} \left(\int_\Omega (u - \pi_{\mathcal{M}} u)^p \right)^{\frac{1}{p}} \qquad [4.26]$$

under the constraint

$$\mathcal{C}(\mathbf{M}) = \int_\Omega d = N.$$

The constraint on the complexity is added to avoid the trivial solution where all $(h_i)_{i=1,3}$ are zero, which provides a null error. Contrary to a discrete analysis, this problem can be solved globally by using a calculus of variations that is well defined on the space of continuous meshes. In the following, to simplify the notations, we define $\gamma_i = {}^t\mathbf{v}_i(\mathbf{x})\, |H_u(\mathbf{x})|\, \mathbf{v}_i(\mathbf{x})$ for $i = 1, 3$. These positive scalar quantities measure the alignment between H_u and the directions of the continuous mesh given by $(\mathbf{v}_i)_{i=1,3}$.

Optimization problem [4.26] is solved in two steps. First, in section 4.4.1, problem [4.26] is solved for the subset of continuous meshes having the same fixed $(\gamma_i)_{i=1,3}$; in other words, we seek for the $(h_i)_{i=1,3}$ solution of [4.26]. Then, in section 4.4.3, the optimal directions $(\mathbf{v}_i)_{i=1,3}$ are derived.

4.4.1. *Formal resolution*

In this section, problem [4.26] is solved for the subset of continuous meshes having the same fixed $(\gamma_i)_{i=1,3}$. The resolution is based on a one-to-one change of variables using the decomposition of proposition 3.2. The density d and the first two anisotropic quotients $(r_i)_{i=1,2}$ are considered instead of the original unknowns $(h_i)_{i=1,3}$. The one-to-one mapping is then given by

$$h_i = d^{-\frac{1}{3}} r_i^{\frac{1}{3}} \text{ for } i = 1,2 \text{ and } h_3 = (r_1\, r_2\, d)^{-\frac{1}{3}}.$$

With this new set of unknowns, function $e_{\mathcal{M}}$ locally writes as follows:

$$e_{\mathcal{M}} = \frac{1}{10}\, d^{-\frac{2}{3}} \left(r_1^{\frac{2}{3}} \gamma_1 + r_2^{\frac{2}{3}} \gamma_2 + (r_1\, r_2)^{-\frac{2}{3}} \gamma_3 \right); \qquad [4.27]$$

thus, we have to solve

$$\min_{((r_i)_i, d)} \int_\Omega d^{-\frac{2p}{3}} \left(r_1^{\frac{2}{3}} \gamma_1 + r_2^{\frac{2}{3}} \gamma_2 + (r_1\, r_2)^{-\frac{2}{3}} \gamma_3 \right)^p$$

under the linear constraint

$$\int_\Omega d = N. \qquad [4.28]$$

One main consequence of considering d as an unknown is to have now a linear constraint so that problem [4.26] becomes convex. This one-to-one mapping also leads to an uncoupled problem: the optimal anisotropic quotients $(r_i)_{i=1,2}$ are first exhibited, and the optimal density is derived in a second step.

The classical Euler–Lagrange necessary condition states that the variation of E_p at point $\mathbf{M}$ in direction $\delta\mathbf{M}$ is proportional to the variation of the constraint $\mathcal{C}$ in the neighborhood of a critical point. In our formal approach, the variation of E_p is approximated by

$$\delta E_p(\mathbf{M}; \delta\mathbf{M}) = \lim_{\varepsilon \to 0} \frac{1}{\varepsilon} \left(\int_\Omega e^p_{\mathcal{M}+\varepsilon\delta\mathcal{M}} - \int_\Omega e^p_{\mathcal{M}} \right) \approx \int_\Omega \frac{\partial e^p_{\mathcal{M}}}{\partial \mathcal{M}}\, \delta\mathcal{M}. \qquad [4.29]$$

As we have an equality constraint, the variation of $\mathcal{C}$ is null so that the necessary Euler–Lagrange condition simplifies to $\delta E_p(\mathbf{M}; \delta\mathbf{M}) = 0$ and $\delta\mathcal{C}(\mathbf{M}; \delta\mathbf{M}) = 0$ for all $\delta\mathbf{M}$. For the variation $\delta\mathbf{M} = ((\delta r_i)_{i=1,2}, \delta d)$, we get

$$\forall \delta r_1,\ \forall \delta r_2,\ \forall \delta d \text{ with } \int \delta d = 0 \text{ we have } \sum_{i=1}^{2} \delta E_p(\mathbf{M}; \delta r_i) + \delta E_p(\mathbf{M}; \delta d) = 0. \qquad [4.30]$$

If Γ stands for $(r_1^{\frac{2}{3}}\gamma_1 + r_2^{\frac{2}{3}}\gamma_2 + (r_1\,r_2)^{-\frac{2}{3}}\gamma_3)$, [4.30] leads to

$$\sum_{i=1}^{2}\delta E_p(\mathbf{M};\delta r_i) = \int_\Omega \frac{2p}{3} d^{-\frac{2p}{3}}(\Gamma)^{p-1}\left(\sum_{i=1}^{2}\left(r_i^{-\frac{1}{3}}\gamma_i - (r_1\,r_2)^{-\frac{2}{3}} r_i^{-1}\gamma_3\right)\delta r_i\right) = 0, \quad [4.31]$$

with the legal choice $\delta d = 0$. A particular condition to ensure [4.31] is given by

$$\begin{cases} r_1^{-\frac{1}{3}}\,\gamma_1 = (r_1\,r_2)^{-\frac{2}{3}}\,r_1^{-1}\gamma_3, \\ r_2^{-\frac{1}{3}}\,\gamma_2 = (r_1\,r_2)^{-\frac{2}{3}}\,r_2^{-1}\gamma_3, \end{cases} \Longrightarrow (r_1\,r_2)^{-1} = \frac{(\gamma_1\,\gamma_2)^{\frac{1}{2}}}{\gamma_3} \Longrightarrow \begin{cases} r_1 = \gamma_1^{-1}\,\gamma_2^{\frac{1}{2}}\,\gamma_3^{\frac{1}{2}}, \\ r_2 = \gamma_1^{\frac{1}{2}}\,\gamma_2^{-1}\,\gamma_3^{\frac{1}{2}}. \end{cases}$$

Previous equalities are used to simplify the initial expression of $e_{\mathcal{M}}$ given by [4.27]:

$$e_{\mathcal{M}} = d^{-\frac{2}{3}}\left(r_1^{\frac{2}{3}}\gamma_1 + r_2^{\frac{2}{3}}\gamma_2 + (r_1\,r_2)^{-\frac{2}{3}}\gamma_3\right) = 3\,d^{-\frac{2}{3}}\,(\gamma_1\,\gamma_2\,\gamma_3)^{\frac{1}{3}}\,.$$

Relation [4.30] for the legal choices $\delta r_i = 0$ for $i = 1, 2$ leads to

$$\delta E_p(\mathbf{M};\delta d) = \int_\Omega \frac{2p}{3} 3^p\, d^{-\frac{2p+3}{3}}\,(\gamma_1\,\gamma_2\,\gamma_3)^{\frac{p}{3}}\,\delta d \;= 0 \;\text{ with }\; \int_\Omega \delta d = 0. \quad [4.32]$$

A condition to ensure [4.32] is given by $d^{-\frac{2p+3}{3}}\,(\gamma_1\,\gamma_2\,\gamma_3)^{\frac{p}{3}} = Cte$, where Cte is a real constant. Using the constraint on the complexity defined by [4.28], the final expression of d is

$$d = N\left(\int_\Omega (\gamma_1\,\gamma_2\,\gamma_3)^{\frac{p}{2p+3}}\right)^{-1}(\gamma_1\,\gamma_2\,\gamma_3)^{\frac{p}{2p+3}}\,.$$

Finally, the optimal continuous mesh $\mathbf{M}^*$ solution of [4.26] is given by

$$\boxed{\begin{aligned} &\lambda_i^* = (h_i^*)^{-2} = N^{\frac{2}{3}}\left(\int_\Omega (\gamma_1\,\gamma_2\,\gamma_3)^{\frac{p}{2p+3}}\right)^{-\frac{2}{3}}(\gamma_1\,\gamma_2\,\gamma_3)^{-\frac{1}{2p+3}}\,\gamma_i, \\ &\text{and }\; r_i^* = \frac{(\gamma_1\,\gamma_2\,\gamma_3)^{\frac{1}{2}}}{\gamma_i^{\frac{3}{2}}} \quad \forall i = 1\ldots 3. \end{aligned}} \quad [4.33]$$

4.4.2. *Uniqueness*

We now prove that the optimal continuous mesh defined by [4.33] is the unique solution of problem [4.26] verifying $E_p(\mathbf{M}^*)^p \leq E_p(\mathbf{M})^p$ for all $\mathbf{M}$ having the same fixed $(\gamma_i)_{i=1,3}$. We can evaluate E_p to the power p at the optimal point $\mathbf{M}^*$:

$$E_p(\mathbf{M}^*)^p = 3^p N^{-\frac{2p}{3}} \left(\int_\Omega (\gamma_1\, \gamma_2\, \gamma_3)^{\frac{p}{2p+3}} \right)^{\frac{2p+3}{3}}. \qquad [4.34]$$

A generic $\mathbf{M}$ of complexity N is given by its three anisotropic quotients $(r_i)_{i=1,3}$ and its density d. To take into account the constraint on the density, the density is rewritten as $d = N(\int_\Omega f)^{-1} f$, where f is a strictly positive function. The error committed with $\mathbf{M}$ is then

$$E_p(\mathbf{M})^p = N^{-\frac{2p}{3}} \left(\int_\Omega f \right)^{-\frac{2p}{3}} \int_\Omega f^{-\frac{2p}{3}} \left(r_1^{\frac{2}{3}} \gamma_1 + r_2^{\frac{2}{3}} \gamma_2 + r_3^{\frac{2}{3}} \gamma_3 \right)^p. \qquad [4.35]$$

To prove $E_p(\mathbf{M}^*)^p \leq E_p(\mathbf{M})^p$, we use the generalized arithmetic-geometric inequality that comes from the concavity of ln:

$$\begin{aligned} \ln\left(\frac{1}{3} \sum_{i=1}^{3} r_i^{\frac{2}{3}} \gamma_i \right) \geq \frac{1}{3} \sum_{i=1}^{3} \ln\left(r_i^{\frac{2}{3}} \gamma_i \right) = \ln\left(\prod_{i=1}^{3} \gamma_i^{\frac{1}{3}} \right) \Longrightarrow \left(\sum_{i=1}^{3} r_i^{\frac{2}{3}} \gamma_i \right)^p \\ \geq 3^p \left(\prod_{i=1}^{3} \gamma_i^{\frac{1}{3}} \right)^p, \end{aligned}$$

as $r_1\, r_2\, r_3 = 1$. Finally, if we denote $g = (\gamma_1\, \gamma_2\, \gamma_3\,)^{\frac{p}{2p+3}}$, we have

$$\begin{cases} E(\mathbf{M}^*)^{\frac{3p}{2p+3}} = 3^{\frac{3p}{2p+3}} N^{-\frac{2p}{2p+3}} \displaystyle\int_\Omega g, \\ E(\mathbf{M})^{\frac{3p}{2p+3}} \geq 3^{\frac{3p}{2p+3}} N^{-\frac{2p}{2p+3}} \left(\displaystyle\int_\Omega f \right)^{\frac{2p}{2p+3}} \left(\displaystyle\int_\Omega f^{-\frac{2p}{3}} g^{\frac{2p+3}{3}} \right)^{\frac{3}{2p+3}}. \end{cases}$$

Using the Hölder inequality:

$$\left(\int_\Omega f\right)^{\frac{2p}{2p+3}} \left(\int_\Omega f^{-\frac{2p}{3}} g^{\frac{2p+3}{3}}\right)^{\frac{3}{2p+3}} = \left(\int_\Omega f\right)^{\frac{2p}{2p+3}} \left(\int_\Omega \left(\frac{g}{f^{\frac{2p}{2p+3}}}\right)^{\frac{2p+3}{3}}\right)^{\frac{3}{2p+3}} \quad [4.36]$$

$$\geq \left(\int_\Omega f^{\frac{2p}{2p+3}} \left(\frac{g}{f^{\frac{2p}{2p+3}}}\right)\right) = \int_\Omega g,$$

as $\frac{2p+3}{2p} \geq 1$, $\frac{2p+3}{3} \geq 1$, and $\frac{2p}{2p+3} + \frac{3}{2p+3} = 1$. Relation [4.36] implies $E_p(\mathbf{M}^*) \leq E_p(\mathbf{M})$ for all $\mathbf{M}$ having the same fixed $(\gamma_i)_{i=1,3}$. As problem [4.26] is strictly convex, the optimal solution $\mathbf{M}^*$ is unique.

4.4.3. *Optimal orientations and main result*

For a given set of directions $(\mathbf{v}_i)_{i=1,3}$, the optimal interpolation error reads

$$\boxed{\|u - \pi_{\mathcal{M}^*} u\|_{\mathbf{L}^p(\Omega)} = 3N^{-\frac{2}{3}} \left(\int_\Omega (\det(\Gamma))^{\frac{p}{2p+3}}\right)^{\frac{2p+3}{3p}},}$$

where Γ is the diagonal matrix composed of $(\gamma_i)_{i=1,3}$. It is then possible to minimize the previous error by seeking the directions $\mathbf{v}_i$ minimizing $\det(\Gamma)^{\frac{p}{2p+3}}$ or equivalently minimizing $\det(\Gamma)$. Geometrically, $\det(\Gamma)$ is simply the volume of the parallelepiped defined by $(\mathbf{v}_i)_{i=1,3}$ and of length $({}^t\mathbf{v}_i\,|H_u|\mathbf{v}_i)_{i=1,3}$. We denote by $(\lambda_i)_{i=1,3}$ the eigenvalues of $|H_u|$ and $(\mathbf{u}_i)_{i=1,3}$ its principal directions. The length of a side computed with respect to $|H_u|$ is

$${}^t\mathbf{v}_1\,|H_u|\,\mathbf{v}_1 = \sum_{k=1}^{3} \lambda_k \left({}^t\mathbf{v}_1\,\mathbf{u}_k\right)^2 \geq \min_{k=1,3} (|\lambda_k|).$$

The previous length is minimal when $\mathbf{v}_1$ is aligned with the eigenvector corresponding to the smallest eigenvalues. We can repeat this procedure among the remaining directions by always aligning the remaining $\mathbf{v}_i$ with the direction associated with the remaining smallest eigenvalue. It results that $\det(\Gamma)^{\frac{p}{2p+3}}$ is minimal when the vectors $(\mathbf{v}_i)_{i=1,3}$ are aligned with the principal directions of $|H_u|$. We conclude that the optimal metric both in sizes and directions is given by the following theorem:

THEOREM 4.3.– Let u be a twice continuously differentiable function defined on $\Omega \subset \mathbb{R}^3$, H_u its Hessian. The optimal continuous mesh $\mathbf{M}_{\mathbf{L}^p}(u)$ minimizing problem [4.26] reads:

$$\boxed{\mathcal{M}_{\mathbf{L}^p} = D_{\mathbf{L}^p}\ \det(|H_u|)^{\frac{-1}{2p+3}}\ |H_u|,\ \text{with}\ D_{\mathbf{L}^p} = N^{\frac{2}{3}} \left(\int_\Omega \det(|H_u|)^{\frac{p}{2p+3}} \right)^{-\frac{2}{3}}.} \qquad [4.37]$$

It verifies the following properties:

– $\mathbf{M}_{\mathbf{L}^p}(u)$ is unique;

– $\mathbf{M}_{\mathbf{L}^p}(u)$ is locally aligned with the eigenvectors basis of H_u and has the same anisotropic quotients as H_u;

– $\mathbf{M}_{\mathbf{L}^p}(u)$ provides an optimal explicit bound of the interpolation error in $\mathbf{L}^p$ norm:

$$\|u - \pi_{\mathcal{M}_{\mathbf{L}^p}} u\|_{\mathbf{L}^p(\Omega)} = 3\, N^{-\frac{2}{3}} \left(\int_\Omega \det\left(|H_u|\right)^{\frac{p}{2p+3}} \right)^{\frac{2p+3}{3p}}. \qquad [4.38]$$

– for a sequence of continuous meshes having an increasing complexity with the same orientation and anisotropic quotients $(\mathbf{M}^N_{\mathbf{L}^p}(u))_{N=1\ldots\infty}$, the asymptotic order of convergence verifies

$$\|u - \pi_{\mathcal{M}^N_{\mathbf{L}^p}} u\|_{\mathbf{L}^p(\Omega)} \leq \frac{Cst}{N^{2/3}}. \qquad [4.39]$$

Relation [4.39] points out a global second order of mesh convergence.

Note that all the constants of [4.38] are explicitly given. In addition, a second order of convergence is predicted. Note also that the family of optimal continuous meshes parameterized by N defines a class of embedded continuous meshes in the sense of definition 3.14. The present method enables to practically generate a discrete mesh approximating the continuous optimal solution by using any metric-based adaptive mesh generator as soon as the generated mesh satisfies definition 2.7. Note that passing to the limit for $p \to \infty$ leads to the classical metric that controls the interpolation error in $\mathbf{L}^\infty$ as used in Castro-Díaz et al. (1997); Frey and Alauzet (2005). We now provide bounds on the discrete interpolation error when the continuous mesh is projected into the space of discrete meshes by means of an adaptive mesh generator. If the mesh generator achieves a unit mesh with respect to $\mathbf{M}_{\mathbf{L}^p}$, then we have the following bounds for the discrete interpolation error:

$$\frac{1}{2}\,\|u-\pi_{\mathcal{M}_{\mathbf{L}^p}}u\|_{\mathbf{L}^p(\Omega)} \leq \|u-\Pi_h u\|_{\mathbf{L}^p(\Omega_h)} \leq 2\,\|u-\pi_{\mathcal{M}_{\mathbf{L}^p}}u\|_{\mathbf{L}^p(\Omega)},$$

or equivalently

$$\begin{aligned}\frac{3}{2}\,N^{-\frac{2}{3}}\left(\int_\Omega \det\left(|H_u|\right)^{\frac{p}{2p+3}}\right)^{\frac{2p+3}{3p}} &\leq \|u-\Pi_h u\|_{\mathbf{L}^p(\Omega_h)}\\ &\leq 6\,N^{-\frac{2}{3}}\left(\int_\Omega \det\left(|H_u|\right)^{\frac{p}{2p+3}}\right)^{\frac{2p+3}{3p}}.\end{aligned}$$

In other words, $\mathbf{M}_{\mathbf{L}^p}$ defined by relation [4.37] allows us to generate an optimal adapted mesh to control the interpolation error in $\mathbf{L}^p$ norm. However, the prescription of the anisotropy is natural and does not need to be transformed into a metric tensors field as done for a posteriori estimates (Picasso 2003) to be able to practically generate quality anisotropic meshes. Lastly, one advantage is that this approach is completely generic and does not depend on the PDE at hand. Gathering the results for space dimension $d = 1, 2, 3$, we get the following optimal metric:

COROLLARY 4.2.– We have $\mathcal{M}_{\mathbf{L}^p} = \mathcal{K}_p(H_u)$ with

$$\mathcal{K}_p(H_u) = D_{\mathbf{L}^p}\,\det(|H_u|)^{\frac{-1}{2p+d}}\,|H_u|,\ \text{and}\ D_{\mathbf{L}^p} = N^{\frac{2}{d}}\left(\int_\Omega \det(|H_u|)^{\frac{p}{2p+d}}\right)^{-\frac{2}{d}}.\ \square \qquad [4.40]$$

REMARK 4.6.– It can be useful to prescribe a target error instead of a target number of vertices. From relation [4.40], the optimal resulting error can be evaluated:

$$\mathcal{E}(\mathcal{M}_{\mathbf{L}^p}) = d\,N^{-\frac{2}{d}}\left(\int_\Omega (\det|H_u|)^{\frac{p}{2p+d}}\right)^{\frac{2p+d}{dp}}. \qquad [4.41]$$

We then deduce the expression of the optimal metric respecting a given error threshold ε:

$$\mathcal{M}_{\mathbf{L}^p} = D_{\mathbf{L}^p}\,\det(|H_u|)^{\frac{-1}{2p+d}}\,|H_u|,\ \text{with}\ D_{\mathbf{L}^p} = \frac{d}{\varepsilon}\left(\int_\Omega \det(|H_u|)^{\frac{p}{2p+d}}\right)^{\frac{1}{p}}.\ \square[4.42]$$

4.5. Multidimensional discontinuity capturing

The analysis of the ability in discontinuity capturing of the 3D multiscale method can be done by investigating when a discontinuity allows the following integral:

$$\|u - \pi_{\mathcal{M}_{\mathbf{L}^p}} u\|_{\mathbf{L}^p(\Omega)} = 3\, N^{-\frac{2}{3}} \left(\int_\Omega \det\left(|H_u|\right)^{\frac{p}{2p+3}} \right)^{\frac{2p+3}{3p}} . \quad [4.43]$$

to be bounded. Let us consider the case of a Heaviside function in a bounded computational subdomain Ω of $I\!R^3$: the discontinuity of u in a plan $x = 0$ of the x, y, z space separating two half space where u is, respectively, 0 and 1. For $x \neq 0$ u, has a zero Hessian and for $x = 0$ u_{xx} is infinitely large while u_{yy}, u_{yz}, u_{zz} are bounded. We regularize by replacing the first and second derivatives of u by divided differences:

$$\begin{aligned} u_{\delta,x}(x,y,z) &= \delta^{-1}(u(x+\delta) - u(x-\delta)), \\ u_{\delta,xx}(x,y,z) &= \delta^{-2}(u(x+\delta) - 2u(x) + u(x-\delta)) \end{aligned}$$

with mixed derivatives

$$u_{\delta,xy}(x,y,z) = \delta^{-1}(u_y(x+\delta) - u_y(x-\delta)) = 0,$$

where δ is assumed to be sufficiently small. Derivatives with respect to y, z are usual derivatives. In a triple row of elements around the discontinuity the approximate second x-derivative is in $O(\delta^{-2})$, the rest are zero. The approximate Hessian of u reduces to the diagonal in Cartesian directions:

$$H_{\delta,u} = \begin{bmatrix} K\delta^{-2} & 0 & 0 \\ 0 & 0 & 0 \\ 0 & 0 & 0 \end{bmatrix} . \quad [4.44]$$

Since Ω is bounded and since the support of $H_{\delta,u}$ is inside $]-2\delta, +2\delta[$, we have

$$\int_\Omega \det\left(|H_{\delta,u}|\right)^\gamma < +\infty \ \text{ for } \gamma \leq \frac{1}{2}. \quad [4.45]$$

Now $p/2p+3$ is always less than $1/2$. Then the RHS of [4.43] is always bounded and *the Heaviside function is approximated by the multiscale adaptation at order* 2 *for any p.*

4.6. Linear interpolate operator

We first introduce some notations. Let Ω_h be a discretization of domain Ω of $\mathbb{R}^d$ and $\mathcal{H}$ a mesh of Ω_h composed of N vertices. The discretized domain Ω_h is assumed to be identical to Ω and thus denoted Ω in the following. We denote by $\{\mathbf{p}_i\}_{i=1..N}$ the set of vertices of mesh $\mathcal{H}$. The approximation space V_h is the space of continuous functions, which are piecewise linear (their restriction to each element is linear):

$$V_h = \{\psi \in \mathcal{C}^0(\Omega) \mid \psi|_K \text{ is affine } \forall K \in \mathcal{H}\}\,.$$

V_h is also called the usual P_1 finite element space. The dimension of V_h is equal to the number of mesh vertices, here N. A natural basis of V_h is the one formed by the basis functions $(\psi_i)_{1\leq i\leq N}$. ψ_i is the function of V_h such that

$$\psi_i(\mathbf{p}_j) = \delta ij \quad \text{for } \mathbf{p}_j \in \mathcal{H} \quad \Longleftrightarrow \quad \begin{cases} \psi_i(\mathbf{p}_i) &= 1 \\ \psi_i(\mathbf{p}_j) &= 0 \quad \text{if } j \neq i \end{cases}.$$

The *patch of function* ψ_i is the set of triangles sharing vertex $\mathbf{p}_i$. These basis functions are also called P_1 *hat functions*. In this basis, any function v_h of V_h reads:

$$v_h(\mathbf{x}) = \sum_{i=1}^{n} v(\mathbf{p}_i)\,\psi_i(\mathbf{x}).$$

We call *shape functions* the restriction of the basis functions to an element. The linear interpolate of a continuous function u on mesh $\mathcal{H}$ is given by:

$$\forall \mathbf{x} \in \Omega,\ \Pi_h u(\mathbf{x}) = \sum_{i=1}^{n} u(\mathbf{p}_i)\,\psi_i(\mathbf{x}),$$

where ψ_i is the ith shape function. We have

$$\forall \mathbf{p}_i \in \mathcal{H},\ \Pi_h u(\mathbf{p}_i) = u(\mathbf{p}_i).$$

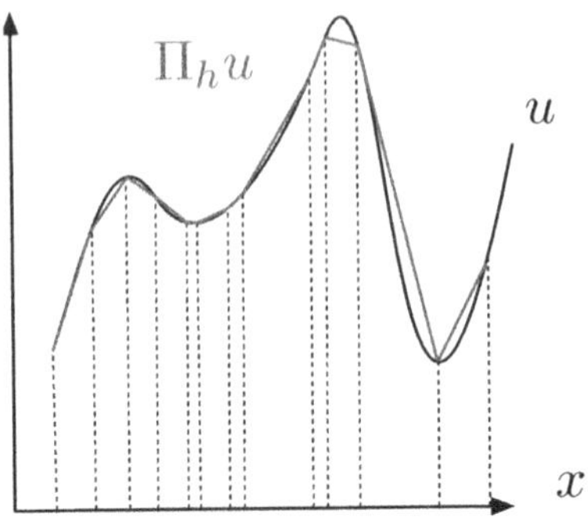

Figure 4.9. *Illustration of the linear interpolate $\Pi_h u$ of a function u. For a color version of this figure, see www.iste.co.uk/dervieux/meshadaptation1*

4.7. A local $\mathbf{L}^\infty$ upper bound of the interpolation error

THEOREM 4.4.– Let K be a tetrahedron of mesh $\mathcal{H}$ defined by its list of edges $(\mathbf{e_j})_{j=1\ldots6}$. Let u be a smooth function two times differentiable and H_u its Hessian matrix. Then, we have the following bound of the interpolation error in $\mathbf{L}^\infty$ norm:

$$\begin{aligned} \| u - \Pi_h u \|_{\mathbf{L}^\infty(K)} &\leq \frac{9}{32} \max_{\mathbf{x},\mathbf{y},\mathbf{z}\in K} \langle \mathbf{yz},\, |H_u(\mathbf{x})|\, \mathbf{yz} \rangle \\ &\leq \frac{9}{32} \max_{\mathbf{x}\in K} \max_{j=1\ldots6} \langle \mathbf{e}_j,\, |H_u(\mathbf{x})|\, \mathbf{e}_j \rangle , \end{aligned}$$

where $|H_u| = \mathcal{R}\,|\Lambda|\,\mathcal{R}^{-1}$ with $|\Lambda| = diag(|\lambda_i|)$. In other words, a bound of the interpolation error in $\mathbf{L}^\infty$ norm is given by the square length of the largest edge of the tetrahedron computed with respect to the metric of the maximal absolute value of the Hessian:

$$\| u - \Pi_h u \|_{\mathbf{L}^\infty(K)} \leq \frac{9}{32} \max_{\mathbf{x}\in K} \max_{j=1\ldots6} \ell^2_{|H_u(\mathbf{x})|}(\mathbf{e}_j). \qquad [4.46]$$

PROOF.– To demonstrate these estimates, we follow the analysis suggested by Anglada et al. (1999) on surfaces. We consider a tetrahedron K. We note $[\mathbf{a}, \mathbf{b}, \mathbf{c}, \mathbf{d}]$ its four vertices and $(\mathbf{e_j})_{j=1\ldots6}$ its six edges. Let u be a smooth function two times differentiable and H_u its Hessian matrix. Moreover, we assume that the Hessian of u is definite, that is, $H_u \neq 0$. Otherwise, u is linear and thus there is no interpolation error. □

The aim is now to bound the linear interpolation error $e = u - \Pi_h u$ on K. We consider a Taylor expansion with integral rest of error function e at one of the vertices of K, for example, $\mathbf{a}$, with respect to any interior point $\mathbf{x}$ in K:

$$(u - \Pi_h u)(\mathbf{a}) = (u - \Pi_h u)(\mathbf{x}) + \langle \mathbf{xa}, \nabla(u - \Pi_h u)(\mathbf{x})\rangle + \int_0^1 (1-t)\,\langle \mathbf{ax}, H_u(\mathbf{x} + t\mathbf{xa})\,\mathbf{ax}\rangle\, dt, \quad [4.47]$$

where $\nabla u(\mathbf{x})$ denotes the gradient of variable u at point $\mathbf{x}$. Actually, we assume that the maximal error is achieved at point $\mathbf{x}$, then:

$$\nabla(u - \Pi_h u)(\mathbf{x}) = 0$$

and this is equivalent to

$$\langle \mathbf{v}, \nabla(u - \Pi_h u)(\mathbf{x})\rangle = 0, \quad \forall \mathbf{v} \subset K.$$

Furthermore, by definition of the linear interpolate we have $u(\mathbf{a}) = \Pi_h u(\mathbf{a})$. Thus, relation [4.47] simplifies to

$$|e(\mathbf{x})| = |(u - \Pi_h u)(\mathbf{x})| = |\int_0^1 (1-t)\,\langle \mathbf{ax}, H_u(\mathbf{x} + t\mathbf{xa})\,\mathbf{ax}\rangle\, dt\,|.$$

Without loss of generality, we assume that $\mathbf{x}$ is closer to $\mathbf{a}$ than to $\mathbf{b}$, $\mathbf{c}$ or $\mathbf{d}$. Let $\mathbf{a}'$ represents the point corresponding to the intersection of line $\mathbf{ax}$ with the face opposite to $\mathbf{a}$. It exists a real number λ such that $\mathbf{ax} = \lambda\,\mathbf{aa}'$. As $\mathbf{a}$ is closer to $\mathbf{x}$ than any other vertex of K, then $\lambda \leq 3/4$. This leads us to write

$$\begin{aligned} |e(\mathbf{x})| &= |\int_0^1 (1-t)\,\lambda^2\,\langle \mathbf{aa}', H_u(\mathbf{a} + t\mathbf{xa})\,\mathbf{aa}'\rangle\, dt\,|, \\ &\leq \frac{9}{16} \max_{\mathbf{y} \in \mathbf{aa}'} |\langle \mathbf{aa}', H_u(\mathbf{y})\,\mathbf{aa}'\rangle\,|\;|\int_0^1 (1-t)\, dt\,| \end{aligned}$$

and then

$$|e(\mathbf{x})| \leq \frac{9}{32} \max_{\mathbf{y} \in K} |\langle \mathbf{aa}', H_u(\mathbf{y})\,\mathbf{aa}'\rangle\,|.$$

Finally, we can introduce the $\mathbf{L}^\infty$-norm of the interpolation error:

$$\|u - \Pi_h u\|_{\mathbf{L}^\infty(K)} \leq \frac{9}{32} \max_{\mathbf{y}\in K} |\langle \mathbf{aa}', H_u(\mathbf{y})\, \mathbf{aa}' \rangle |.$$

At this point, it can be of practical interest to introduce a metric tensor in the previous relationship. The 3×3 Hessian matrix being symmetric, it admits the spectral decomposition $H_u = \mathcal{R}\,\Lambda\,{}^t\mathcal{R}$ where $\mathcal{R}$ and $\Lambda = diag(\lambda_i)$ are the eigenvectors and the eigenvalues matrix, respectively. We introduce the absolute value of the Hessian defined by $|H_u| = \mathcal{R}\,|\Lambda|\,{}^t\mathcal{R}$ with $|\Lambda| = diag(|\lambda_i|)$. $|H_u|$ is a metric tensor:

$$\begin{aligned}\|u - \Pi_h u\|_{\mathbf{L}^\infty(K)} &\leq \frac{9}{32} \max_{\mathbf{y}\in K} \langle \mathbf{aa}', |H_u(\mathbf{y})|\, \mathbf{aa}' \rangle \\ &\leq \frac{9}{32} \max_{y\in K} \max_{\vec{v}\subset K} \langle \vec{v}, |H_u(y)|\, \vec{v} \rangle.\end{aligned}$$

Note that the previous relationship is not very useful in practice because of the max. However, it can be shown that

$$\forall \mathbf{v} \subset K, \quad \|\mathbf{v}\|_{|H_u(\mathbf{x})|} \leq \max_{j=1\ldots 6} \|\mathbf{e}_j\|_{|H_u(\mathbf{x})|}$$

(see Alauzet and Frey 2003 for the proof). Previous bound can then be reformulated as follows:

$$\|u - \Pi_h u\|_{\mathbf{L}^\infty(K)} \leq \frac{9}{32} \max_{\mathbf{y}\in K} \max_{j=1\ldots 6} \langle \mathbf{e}_j, |H_u(\mathbf{y})|\, \mathbf{e}_j \rangle. \qquad [4.48]$$

This expression provides a bound in the case where the maximum error is reached inside element K. In the case where the maximum error is achieved on a face or on an edge, a similar proof can be done.

If the maximum error is obtained on element face $F = [\mathbf{a}, \mathbf{b}, \mathbf{c}]$ with edges $(\mathbf{e_j})_{j=1\ldots 3}$, then we obtain

$$\|u - \Pi_h u\|_{\mathbf{L}^\infty(K)} \leq \frac{2}{9} \max_{\mathbf{y}\in F} \max_{j=1\ldots 3} \langle \mathbf{e}_j, |H_u(\mathbf{y})|\, \mathbf{e}_j \rangle.$$

Similarly, for a maximum value obtained along edge $\mathbf{e}_j$, we have

$$\|u - \Pi_h u\|_{\mathbf{L}^\infty(K)} \leq \frac{1}{8} \max_{\mathbf{y}\in \mathbf{e}_j} \langle \mathbf{e}_j, |H_u(\mathbf{y})|\, \mathbf{e}_j \rangle.$$

In conclusion, as $\frac{1}{8} < \frac{2}{9} < \frac{9}{32}$, relation [4.48] provides a proper bound for the interpolation error on element K.

To demonstrate relation [4.46], we use the definition of the length in a metric space, see relation [3.1]. We have: $\ell^2_{|H_u(\mathbf{y})|}(\mathbf{e}_j) = \langle \mathbf{e}_j, |H_u(\mathbf{y})|\, \mathbf{e}_j \rangle$. Consequently, the previous bound can be rewritten as

$$\| u - \Pi_h u \|_{\mathbf{L}^\infty(K)} \leq \frac{9}{32} \max_{\mathbf{x}\in K} \max_{j=1\ldots 6} \ell^2_{|H_u(\mathbf{x})|}(\mathbf{e}_j).$$

This ends the proof. □

REMARK 4.7.– The constant appearing in the interpolation error bound can be evaluated for any dimension by induction. In dimension d, we have

$$c_d = \lambda^2 \left| \int_0^1 (1-t)\mathrm{d}t \right| = \frac{1}{2}\left(\frac{d}{d+1}\right)^2.$$

REMARK 4.8.– Theorem 4.4 allows us to estimate the interpolation error on any element K:

$$\varepsilon_K = c_d \max_{\mathbf{y}\in K} \max_{j=1\ldots 6} \langle \mathbf{e}_j, |H_u(\mathbf{y})|\, \mathbf{e}_j \rangle,$$

where ε_K is the estimated error on K.

4.8. Metric construction for mesh adaptation

Let us give a short synthesis of the previous chapters. We have seen that

1) metric tensor $\mathcal{M}$ can be used to prescribe sizes directionally to the mesher;

2) to generate an adapted mesh $\mathcal{H}$, the mesher generates a unit mesh with respect to Riemannian metric space $(\mathcal{M}(\mathbf{x}))_{\mathbf{x}\in\Omega}$;

3) the previous theorem provides an estimate of the interpolation error of function u on an element K.

Now, we aim at generating an anisotropic adapted mesh to control the interpolation error model. How can we link these three items? To this end, we fix the error threshold to ε and we want to commit an error of ε on each element K of mesh $\mathcal{H}$:

$$\forall K \in \mathcal{H},\; c_d \max_{\mathbf{x}\in K} \max_{j=1\ldots 6} \langle \mathbf{e}_j, |H_u(\mathbf{x})|\, \mathbf{e}_j \rangle = c_d \max_{\mathbf{x}\in K} \max_{j=1\ldots 6} \ell^2_{|H_u(\mathbf{x})|}(\mathbf{e}_j) = \varepsilon.$$

However, the left-hand term remains hard to compute numerically due to the $\max_{\mathbf{x}\in K}$ on the Hessian. Indeed, the only information we have on the Hessian is a discrete one: the pointwise values at mesh vertices $\{|H(\mathbf{p}_i)|\}_{\mathbf{p}_i\in\mathcal{H}}$. However, the $\max_{\mathbf{x}\in K}$ operator requires the knowledge of $H_u(\mathbf{x})$ everywhere inside tetrahedron K, that is, a continuous information. Because of the metric interpolation operator, section 3.6.2, we are able to define a Riemannian metric space $(|H(\mathbf{x})|)_{\mathbf{x}\in\Omega}$. We thus propose to control the error by generating a mesh $\mathcal{H}$ such that

$$\forall K \in \mathcal{H},\ \ c_d \max_{j=1\ldots6} \ell^2_{|H_u|}(\mathbf{e}_j) = \varepsilon.$$

This relation points out that it is sufficient to control the length of the mesh edges in Riemannian metric space $(|H(\mathbf{x})|)_{\mathbf{x}\in\Omega}$ to control the interpolation error on each element of the mesh. Our problem of mesh adaptation reduces to generate a mesh $\mathcal{H}$ such that

$$\forall \mathbf{e} \in \mathcal{H},\ \ c_d\, \ell^2_{|H_u|}(\mathbf{e}) = \varepsilon.$$

Notwithstanding, we still have to make the link with the notion of unit mesh. As we have

$$c_d\,\langle \mathbf{e},\, |H_u|\,\mathbf{e}\rangle = \varepsilon \quad\Longleftrightarrow\quad \langle \mathbf{e},\, \frac{c_d\,|H_u|}{\varepsilon}\,\mathbf{e}\rangle = 1,$$

the mesh adaptation problem can finally be formulated as follows:

Generate a unit mesh $\mathcal{H}$ in Riemannian metric space $(\mathcal{M}(\mathbf{x}))_{\mathbf{x}\in\Omega} = (\frac{c_d\,|H_u|}{\varepsilon}(\mathbf{x}))_{\mathbf{x}\in\Omega}$:

$$\forall \mathbf{e} \in \mathcal{H},\ \ \ell^2_{\mathcal{M}}(\mathbf{e}) = 1.$$

This mesh will be optimal to control the interpolation error in $\mathbf{L}^\infty$-norm. The mesh adaptation strategy described above consists of *equidistributing* the interpolation error over the mesh.

REMARK 4.9.– Let us give a concrete geometric interpretation of this result. If one considers a linear function in 1D (a line), then only two points are needed to exactly represent this function. On the other hand, in the case of a curve, the number of points which are required to describe exactly this curve is not finite in general. To describe this curve with a given accuracy, it is mandatory to know its intrinsic properties, notably its local curvature. The points density must be locally proportional to the curvature. Therefore, we need a control based on the second derivative of the function. In other words, we do not adapt to the function slope (its gradient) but to the function curvature (its Hessian). This error estimate measures the maximal gap between the curve and its piecewise linear representation on the mesh. In 2D, the function can be interpreted as a Cartesian surface embedded in $\mathbb{R}^3$ (see Figure 4.10): $\Sigma = (x, y, u(x, y))$. Analogous to the first dimension, the error estimate measures the gap between the underlying surface Σ and its piecewise linear approximation on the mesh defined by $\Pi_h u$. It imposes a point density proportional to the principal curvatures of the surface. In 3D, the function can be perceived as a Cartesian hypersurface embedded in $\mathbb{R}^4$. This is the justification for the term of *geometric error estimate*, indicating also that it is *problem-independent*.

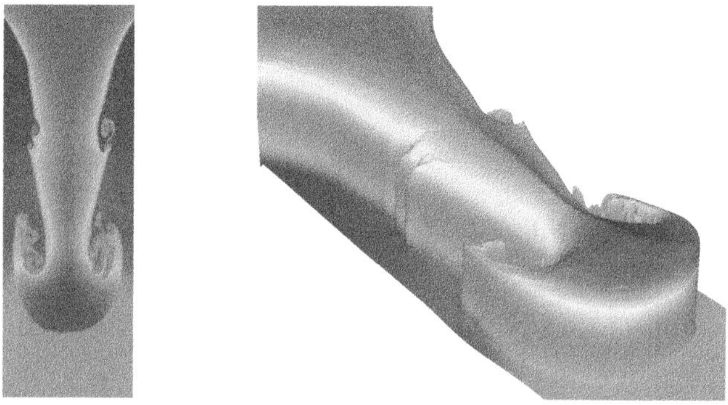

Figure 4.10. *Left, a 2D function; right, its representation as a Cartesian surface in $\mathbb{R}^3$. For a color version of this figure, see www.iste.co.uk/dervieux/meshadaptation1*

REMARK 4.10.– It is important to distinguish the Cartesian surface representing the solution and a manifold that has $(\mathcal{M}(\mathbf{x}))_{\mathbf{x}\in\Omega}$ as natural Riemannian metric (the first fundamental form).

4.8.1. *Handling degenerated cases*

In practice, the Hessian of function u can be 0, for example, if u is linear, then $|H_u|$ is not definite. In this particular case, the interpolation error is 0 and we want to prescribe a mesh size, which is infinite. To solve this issue, this infinite size prescription is truncated by imposing a *maximal size* h_{max}. This is equivalent to truncate tiny eigenvalues by a $\lambda_{min} = h_{max}^{-2}$. The metric is then defined by

$$\mathcal{M} = \mathcal{R}\tilde{\Lambda}^t\mathcal{R} \quad \text{where} \quad \tilde{\Lambda} = diag(\tilde{\lambda}_i), \quad \text{and} \quad \tilde{\lambda}_i = \max\left(\frac{c_d\,|\lambda_i|}{\varepsilon},\, h_{max}^{-2}\right),$$

the $(\lambda_i)_i$ being the eigenvalues of H_u.

Similarly, if one wants to control the overall mesh size of the adapted mesh, a minimal size h_{min} is imposed to truncate to small size prescription. This is equivalent to truncate large eigenvalues by a $\lambda_{max} = h_{min}^{-2}$. The metric is then defined by

$$\mathcal{M} = \mathcal{R}\tilde{\Lambda}^t\mathcal{R} \quad \text{where} \quad \tilde{\Lambda} = diag(\tilde{\lambda}_i), \qquad [4.49]$$

and

$$\tilde{\lambda}_i = \min\left(\max\left(\frac{c_d\,|\lambda_i|}{\varepsilon}, h_{max}^{-2}\right), h_{min}^{-2}\right).$$

4.8.2. *Isotropic mesh adaptation*

Relation [4.49] provides an anisotropic expression of the metric. If an isotropic size prescription is expected, the isotropic metric is obtained by taking into account the smallest size prescription, that is, the largest eigenvalue. This metric is given by

$$\mathcal{M}_{iso} = diag(\max_i(\tilde{\lambda}_i)) = \begin{pmatrix} \max_i(\tilde{\lambda}_i) & 0 & 0 \\ 0 & \max_i(\tilde{\lambda}_i) & 0 \\ 0 & 0 & \max_i(\tilde{\lambda}_i) \end{pmatrix}. \qquad [4.50]$$

4.9. Mesh adaptation for analytical functions

4.9.1. *Algorithms*

Let us discuss an algorithm for finding an adapted mesh of given complexity N for a given function f. For a 2D Heaviside function, the exact Hessian is zero almost everywhere and would imply a uniform "optimal" mesh, and an inaccurate interpolate. We have to consider the Hessian of the interpolate on the current mesh. Different options to do this are studied in Chapter 5. The Hessian of the interpolate depends (nonlinearly) on the mesh, and therefore the algorithm needs to be iterative. We therefore propose the iterative algorithm 4.1. to solve it.

REMARK 4.11 (Complexity or number of nodes).– We want to emphasize that the number of nodes is different but close to the complexity. It is easy to modify the mesh generation algorithm into a loop, which will specify *solely approximatively* the number of nodes/vertices, instead of the complexity. This gives Algorithm 4.2. □

Algorithm 4.1. Mesh adaptation for interpolating an analytical function. This loop provides a couple $(mesh, flow\ solution)$ for a prescribed complexity N according to the Hessian of f.

Initial mesh $\mathcal{H}_0^0$, set error threshold ε.

While i $\leq n_{adap}$ **do**

1) $(f_{i-1}, H_{f_{i-1}}) =$ `Evaluate` f `and its Hessian` H_f `on mesh` $\mathcal{H}_{i-1}$

2) $\mathcal{M}_{i-1} =$ `Compute metric` $\mathcal{M}$ `according to Relation` (4.49)

3) $\mathcal{H}_i =$ `Generate a new N` $-$ `complex adapted mesh from pair` $(\mathcal{H}_{i-1}, \mathcal{M}_{i-1})$

Endwhile

Algorithm 4.2. Mesh adaptation for interpolating an analytical function. This loop provides a couple $(mesh, flow\ solution)$ for a prescribed number of about NS vertices according to the Hessian of f.

Initial mesh $\mathcal{H}_0^0$, set error threshold ε.

While i $\leq n_{adap}$ **do**

1) $(f_{i-1}, H_{f_{i-1}}) =$ `Evaluate` f `and its Hessian` H_f `on mesh` $\mathcal{H}_{i-1}$

2) $\mathcal{M}_{i-1} =$ `Compute metric` $\mathcal{M}$ `according to Relation` (4.49)

3) $\mathcal{H}_i =$ `Generate a new adapted mesh of about` NS `nodes from pair` $(\mathcal{H}_{i-1}, \mathcal{M}_{i-1})$ `by adjusting iteratively the complexity` N

Endwhile

4.9.2. *Examples of L^∞ adaptation*

Example 1: The χ-shaped function. We consider a square domain $\Omega = [0,1] \times [0,1]$ and function f_1 defined by:

$$f_1(x,y) = \tanh\big(-100\,(y - 0.5 - 0.25\sin(2\pi x))\big) + \tanh\,(100\,(y - x))\,.$$

This function is represented in Figure 4.12. We start from a uniform mesh $\mathcal{H}_{unif}$ composed of 5,417 vertices and 10,585 triangles (Figure 4.11, left). We set the error threshold to $\varepsilon = 0.04$ and after six mesh adaptation iterations, we obtain an anisotropic adapted mesh $\mathcal{H}_{adap}$ composed of 5,132 vertices and 10,084 triangles, depicted in Figure 4.11 (right).

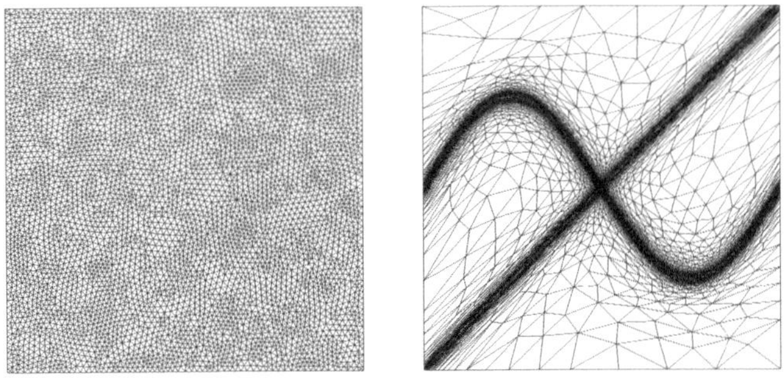

Figure 4.11. *Left: uniform initial mesh; right: final anisotropic adapted mesh for function f_1*

In Figure 4.12, the representation of function f_1 on the adapted mesh (right), that is $\Pi_{\mathcal{H}_{adap}} f_1$, is visually more accurate than its representation on the uniform mesh (left) with the same number of vertices. Mathematically, we can evaluate an approximation of the interpolation error in $\mathbf{L}^\infty$-norm on each mesh, we get

$$\begin{aligned}\|f_1 - \Pi_{\mathcal{H}_{unif}} f_1\|_{\mathbf{L}^\infty(\Omega)} &\approx 4.63 \times 10^{-1}\\ \|f_1 - \Pi_{\mathcal{H}_{adap}} f_1\|_{\mathbf{L}^\infty(\Omega)} &\approx 6.74 \times 10^{-3}.\end{aligned}$$

The interpolation error has been reduced by a factor 70 which confirms our visual intuition.

Example 2: The 5*-wells function.* We consider a square domain $\Omega = [0,1] \times [0,1]$ and function f_2 defined by

$$\begin{aligned} f_2(u,v) &= \tanh\left(30(u^2 + v^2 - \epsilon)\right)\\ &+ \tanh\left(30((u-0.75)^2 + (v-0.75)^2 - \epsilon)\right)\\ &+ \tanh\left(30((u-0.75)^2 + (v+0.75)^2 - \epsilon)\right)\\ &+ \tanh\left(30((u+0.75)^2 + (v-0.75)^2 - \epsilon)\right)\\ &+ \tanh\left(30((u+0.75)^2 + (v+0.75)^2 - \epsilon)\right)\\ &\text{with } \epsilon = 0.25 \quad \text{and} \quad u = 4x-2,\ v = 4y-2. \end{aligned}$$

This function is represented in Figure 4.14. We start from the same uniform mesh $\mathcal{H}_{unif}$ composed of 5,417 vertices and 10,585 triangles (Figure 4.11, left). We set the error threshold to $\varepsilon = 0.15$ and after 6 mesh adaptation iterations, and we obtain an anisotropic adapted mesh $\mathcal{H}_{adap}$ composed of 5,029 vertices and 10,013 triangles, depicted in Figure 4.13 (right).

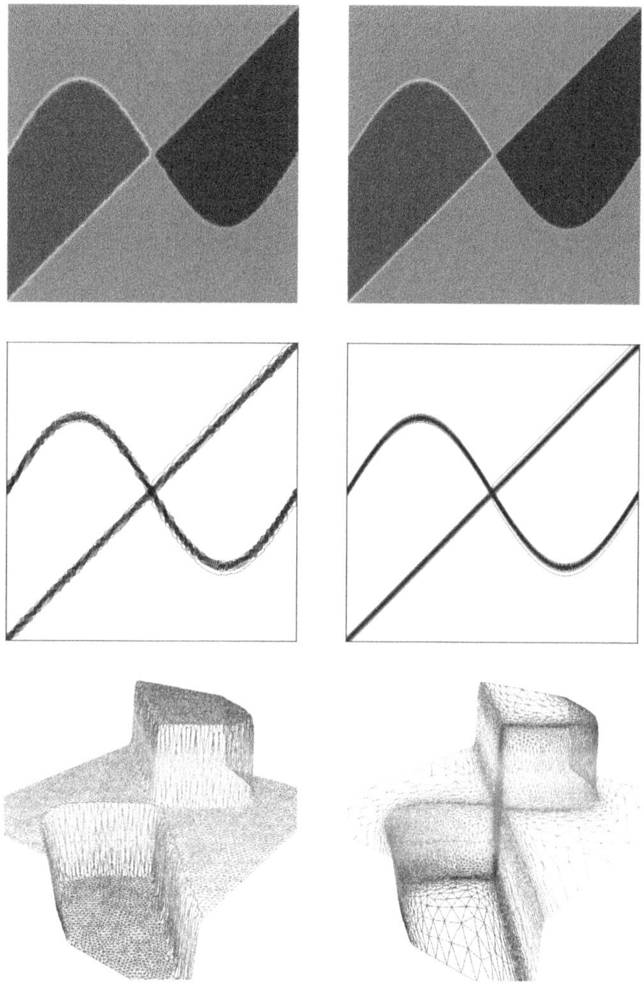

Figure 4.12. *Representation of function f_1 on uniform initial mesh (left) and on final anisotropic adapted mesh (right). For a color version of this figure, see www.iste.co.uk/dervieux/meshadaptation1*

In Figure 4.12, the representation of function f_2 on the adapted mesh (right), that is, $\Pi_{\mathcal{H}_{adap}} f_2$, is visually more accurate that its representation on the uniform mesh (left) with the same number of vertices. Notably, in regions where wells intersect. Mathematically, we can evaluate an approximation of the interpolation error in $\mathbf{L}^\infty$-norm on each mesh, we get:

$$\|f_2 - \Pi_{\mathcal{H}_{unif}} f_2\|_{\mathbf{L}^\infty(\Omega)} \approx 3.40 \times 10^{-1}$$
$$\|f_2 - \Pi_{\mathcal{H}_{adap}} f_2\|_{\mathbf{L}^\infty(\Omega)} \approx 3.43 \times 10^{-2}.$$

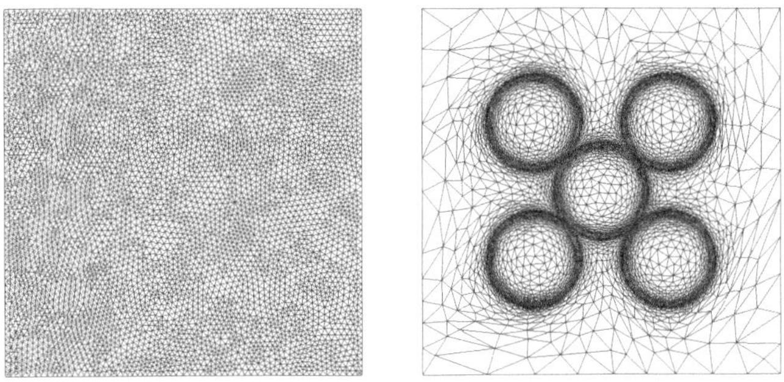

Figure 4.13. *Left: uniform initial mesh; right: final anisotropic adapted mesh for function f_2*

The interpolation error has been reduced by a factor 10, which confirms our visual intuition.

Comparison between isotropic and anisotropic mesh adaptation: To finish with these analytical examples, we compare anisotropic mesh adaptation given by relation [4.49] and isotropic mesh adaptation given by relation [4.50]. Adapted isotropic meshes for both functions contain almost 5,000 vertices and are depicted in Figure 4.15. The interpolation errors are not so different:

$$\|f_1 - \Pi_{\mathcal{H}_{iso}} f_1\|_{\mathbf{L}^\infty(\Omega)} \approx 9.52 \times 10^{-2}$$
$$\|f_2 - \Pi_{\mathcal{H}_{iso}} f_2\|_{\mathbf{L}^\infty(\Omega)} \approx 8.90 \times 10^{-2}.$$

However, a strong improvement is seen in Figure 4.15.

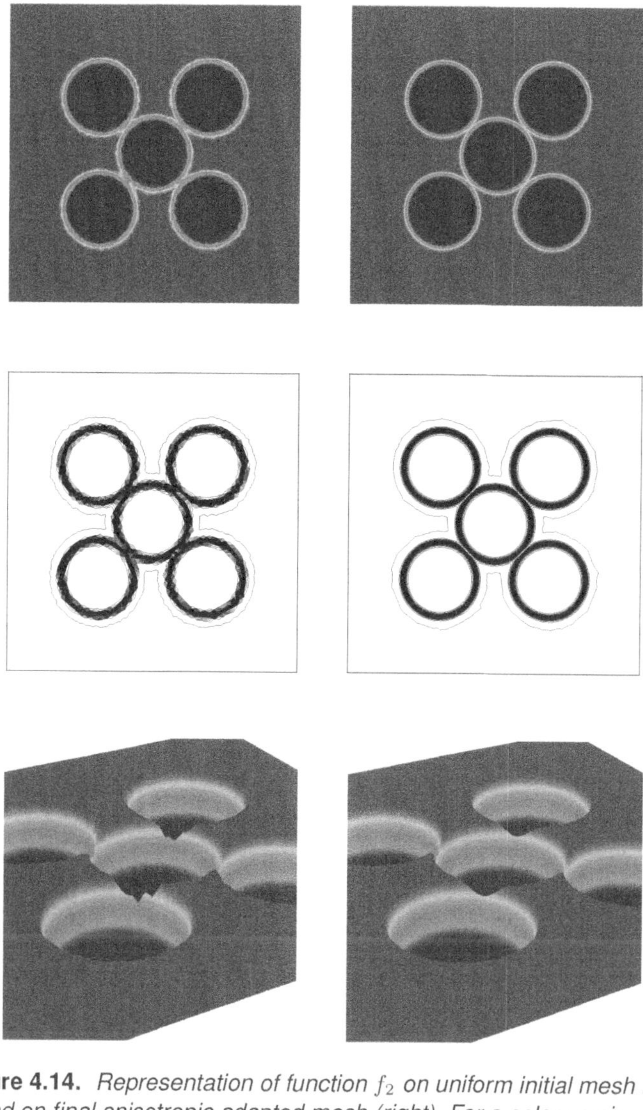

Figure 4.14. *Representation of function f_2 on uniform initial mesh (left) and on final anisotropic adapted mesh (right). For a color version of this figure, see www.iste.co.uk/dervieux/meshadaptation1*

REMARK 4.12 (L^∞ vs L^2: A very sensitive case).– Given a function defined on a fine (uniform or not) mesh, a compression could consist of storing it in a smaller mesh, accepting some loss in the accuracy of its definition, as far as the new file is sufficiently small. Mesh adaptive interpolation is an answer to this problem, already used in image processing (Marquant et al. 2000). We consider the sum of an arctangent function of amplitude 1, combined this time with a (slightly different) sine function of 10 times lower amplitude: $f(x, y) = 0.1 * \sin(50x) + atan(0.001/(\sin(5y) - 2x))$. We compare two adapted meshes with about 2,000 nodes each. The first one is adapted following the error equidistribution principle, in other words, by minimizing the L^∞ error functional. The second one is adapted according to the minimization of the L^2 error norm. We observe that L^2 option restitutes the low amplitude sine oscillation while the L^∞ does not show it at all. □

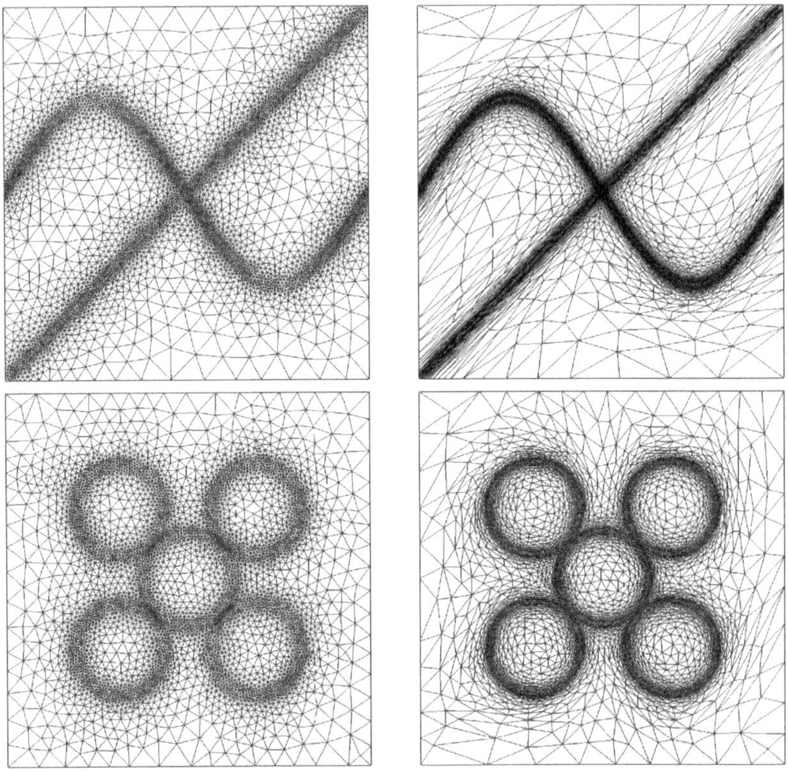

Figure 4.15. *Final isotropic (left) and anisotropic (right) adapted meshes for functions f_1 (top) and f_2 (bottom)*

REMARK 4.13 (Sensitivity of the optimality condition).– The choice of p in the multiscale L^p approach determines solely the optimal mesh density, not the stretching. But this density can be very sensitive to p. We consider the interpolation on the unit disk of $f(x, y) = 10x^3 + y^3 + atan(0.001/(\sin(5y) - 2x))$. Let β be a positive parameter, we consider a series of meshes, indexed by β, that all have about 2,100 nodes and are adapted according to the following formula:

$$\mathcal{M}_{opt} = \left(|\frac{\partial^2 u}{\partial \xi^2}|.|\frac{\partial^2 u}{\partial \eta^2}| \right)^{\beta} \mathcal{R}^{-1} \begin{pmatrix} \left(\frac{|\frac{\partial^2 u}{\partial \xi^2}|}{|\frac{\partial^2 u}{\partial \eta^2}|} \right)^{1/2} & 0 \\ 0 & \left(\frac{|\frac{\partial^2 u}{\partial \eta^2}|}{|\frac{\partial^2 u}{\partial \xi^2}|} \right)^{1/2} \end{pmatrix} \mathcal{R}.$$

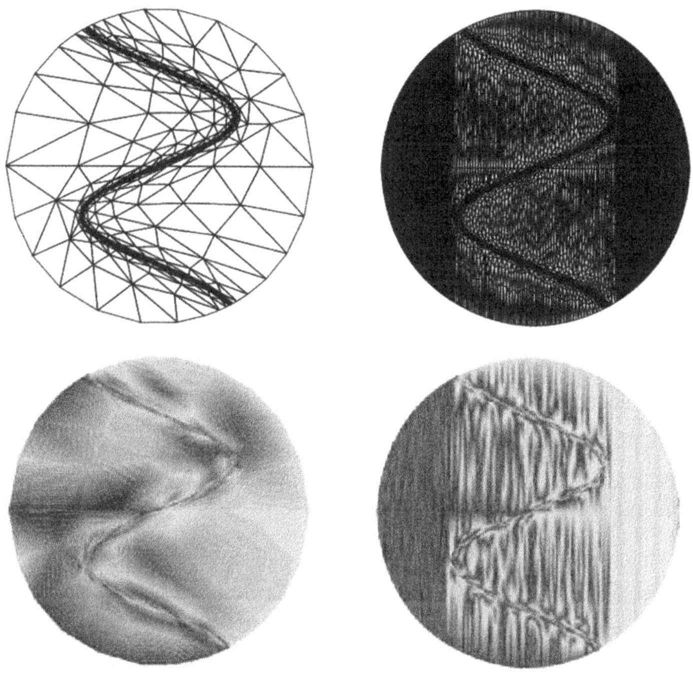

Figure 4.16. *Meshes and contours with the two different options. From left to right, the mesh with L^∞-adaptation , the mesh with L^2-adaptation, the representation with L^∞-adaptation, the representation with L^2-adaptation. The L^∞-adaptation is unable to represent the oscillating behavior of the tested function while L^2-adaptation (right) performs rather well*

This family of metrics involves the usual equidistribution or L^∞ metric, for $\beta = 1$ and the theoretically optimal metric for the L^2 norm, for $\beta = 5/6$. We then compute the L^2 error. Outputs are depicted in Figure 4.17. The error norm related to $\beta = 5/6$

is the lowest one, and is about three times smaller than those resulting from $\beta = 0.7$ or from $\beta = 1$. □

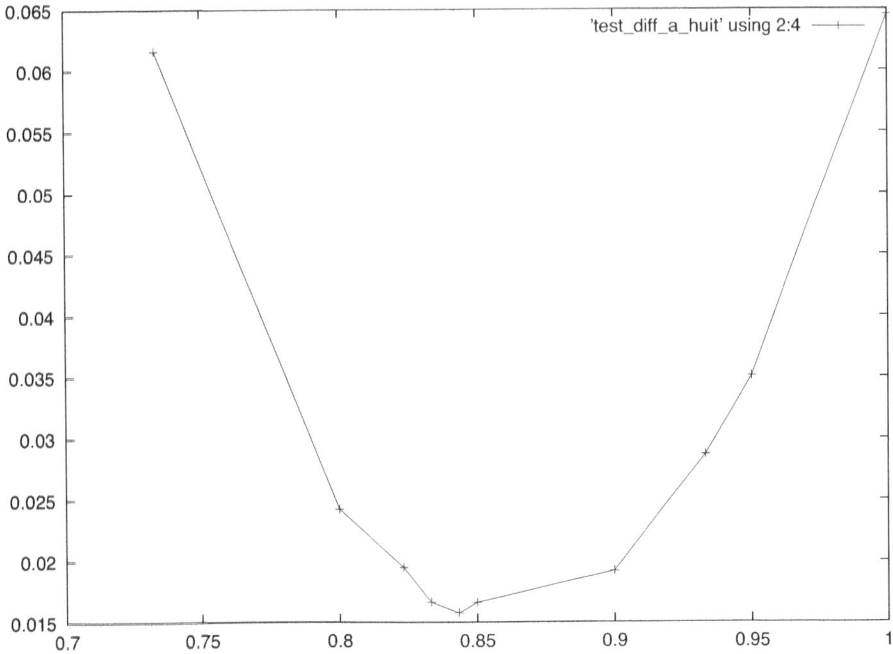

Figure 4.17. *Optimality of the proposed metric: abscissas are values of the β parameter in the adaptation criterion, and ordinates are values of the resulting L^2 interpolation error. The theoretical value of $5/6$ is numerically optimal*

4.10. Conclusion

This chapter has explored some capabilities of a continuous setting for mesh adaptation purposes. This study focalizes on the problem of best adaptive mesh for pure interpolation. A mesh is modeled by a metric. The total number of nodes is a continuous integral of the metric. The continuous interpolation error is modeled with the first term of a Taylor series for the interpolation error. The norm in $L^p, 1 \leq p < \infty$ of the error model is then minimized. We get a completely explicit expression of the optimal metric in terms of the function to which it is adapted. When a discontinuous function is considered, a higher order of accuracy of the adaptation model is predicted for the proposed example. It can help in specifying conditions for practical second-order accuracy. The usual equirepartition strategy or L^∞ analysis appears as a limiting case, which does not enjoy these higher-order convergence properties.

Transposition to the discrete context is demonstrated by a few numerical experiments giving some (yet partial, of course) confirmation that the approach is sound, efficient and shows the behaviors predicted by the theory.

This type of analysis has a potential usefulness for several applications, such as

– *in image compression*: the continuous metric method defines an optimal compression method on an isotropic mesh;

– *in scientific computing*: in the following chapters, we follow the optimal metric strategy proposed here for extending the present method to the research of an optimal mesh for the solution of a PDE.

4.11. Notes

Bound [4.38] has been derived in Chen et al. (2007).

Examples of image compression with the methods developed here can be found in Courty et al. (2006).

5

Multiscale Adaptation for Steady Simulations

Multiscale adaptation has been previously defined for the interpolation of known analytic or continuous functions. The multiscale adaptation for PDE works on the same basis as an exact analytic function, with the difference that information is derived from numerical solutions. In a feature-based approach, we choose a special field, the sensor, generally identical to the solution when it is a scalar field. The derivatives of the sensor need to be computed. This is permitted by *recovery methods*. On the new mesh, we need a good initial solution for recomputing the numerical solution, and this is done by *interpolation methods*.

5.1. Introduction

While previous chapters address the multiscale adaptation to a known function, this chapter focuses on the application of mesh adaptation to numerical simulations. The relation between a numerical solution and the derivation of a better mesh is very frequently based on the transformation of the existing mesh, using information derived from the numerical solution already computed. We give in the notes (section 5.9) a short review of these approaches. In contrast, in the standpoint which is proposed by this chapter, the mesh is defined as *an approximate of the best mesh for the exact solution*.

Then two mains issues have to be addressed:

i) the solution u of the problem is not known, neither are its linear interpolate $\Pi_h u$ and its Hessian H_u. We only know the *numerical solution* u_h and we want to control the *approximation error* $\|u - u_h\|$;

ii) the mesh adaptation process has to be coupled with the flow solver.

Concerning the first issue, the geometric error estimate presented in Chapter 3 cannot be applied directly since u is not available. A complementary treatment is added and complementary assumptions are made to exhibit the metric from the numerical solution u_h. The second point arises once a new adapted mesh has been generated. Indeed, the numerical solution is not known on this new mesh. One can restart the simulation from scratch but this will significantly degrade the efficiency of the method in term of CPU time. To solve this issue, we introduce a new step: the interpolation stage. It consists of projecting the computed solution on this new mesh. Then, the flow solver restarts the simulation from the last state, hence avoiding to waste CPU time. This chapter describes the mesh adaptation scheme in the context of a steady flow computed with triangulations/tetrahedrizations and vertex-centered approximations. Methods for recovering Hessians are described. Interpolations methods are then discussed. A numerical example concludes this chapter.

5.2. Definitions and notations (2D)

Let us consider a bounded domain $\Omega \in \mathbb{R}^2$ and let $\partial\Omega$ be its boundary. We consider a triangular mesh $\mathcal{H} = \bigcup K_i$ of domain Ω composed with triangles. A triangle in $\mathbb{R}^2$ is defined by the list of its vertices and coordinates, which are locally numbered in a convenient way. This list, enriched with some conventions, provides the complete definition of the related element, including the definition of its edges and neighbors, together with an orientation. In particular, the oriented local numbering of the triangle vertices enables us to compute its surface area while giving a sense to its sign. It also enables directional normals to be evaluated for each edge. Formally speaking, the local numberings of vertices, edges and elements are pre-defined in such a way that some properties are implicitly induced. In the case of a triangle with vertices $[\mathbf{p}_0, \mathbf{p}_1, \mathbf{p}_2]$ in this order, the first vertex having been chosen, the numbering of the others is deduced counterclockwise (see Figure 5.1, left). This orientation provides us with positive sign while computing the triangle surface area. Then, the topology can now be well defined because of the edges definition: $\mathbf{e}_0 = \mathbf{p}_1\mathbf{p}_2$, $\mathbf{e}_1 = \mathbf{p}_2\mathbf{p}_0$ and $\mathbf{e}_2 = \mathbf{p}_0\mathbf{p}_1$. This numeration is such that the index of the edge is the index of the viewing vertex, that is, the opposite vertex. Regarding the neighboring triangles, we denote by K_i the neighbor viewing vertex $\mathbf{p}_i$ through edge $\mathbf{e}_i$ (see Figure 5.1, left).

In the following all the indices in square bracket are given modulo 3: $[i] = i\ mod(3)$. Let $K = [\mathbf{p}_0, \mathbf{p}_1, \mathbf{p}_2]$ be a triangle, and its *signed area* A_K is given as follows:

$$A_K = \frac{1}{2}\begin{vmatrix} x_0 & y_0 & 1 \\ x_1 & y_1 & 1 \\ x_2 & y_2 & 1 \end{vmatrix} = \frac{1}{2}\begin{vmatrix} x_1 - x_0 & x_2 - x_0 \\ y_1 - y_0 & y_2 - y_0 \end{vmatrix},$$

where $\mathbf{p}_i = (x_i, y_i)$. This area is *positive* if the triangle is numbered counterclockwise, which is our convention on the mesh orientation. The signed area is also given by one half of the z-component of $\mathbf{p}_0\mathbf{p}_1 \wedge \mathbf{p}_0\mathbf{p}_2$.

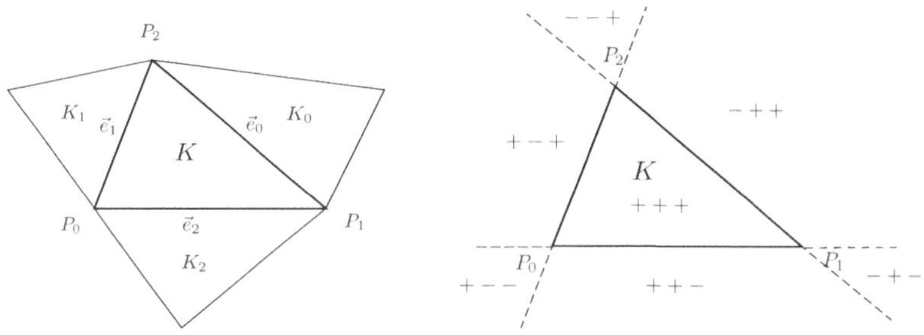

Figure 5.1. *Left: Definition of triangle K and its three neighbors K_i. Vertices indices are ordered counterclockwise and the neighboring entities numeration is the same as the viewing vertices. Right: The seven regions defined by the signs of the three barycentric coordinates of a point $\mathbf{p}$ with respect to element K*

Let $\mathbf{p}$ be a point; we denote by K^i the virtual triangle where vertex $\mathbf{p}_i$ is substituted by $\mathbf{p}$. The signed areas A_{K^i}, for $i = 0 \ldots 2$, are called the *barycentrics* of $\mathbf{p}$. The three associated *barycentric coordinates* are given by

$$\beta_i = \frac{A_{K^i}}{A_K} \quad \text{for} \quad i = 0 \ldots 2.$$

The sign of the three barycentric coordinates or barycentrics defines explicitly seven regions of the plane where point $\mathbf{p}$ can be located with respect to element K (see Figure 5.1, right).

We denote by $\{\mathbf{p}_i\}_{i=1..N}$ the set of vertices of mesh $\mathcal{H}$. The numerical solution u_h lies in the approximation space V_h of the continuous functions, which are piecewise linear:

$$V_h = \{\psi \in \mathcal{C}^0(\Omega) \mid \psi|_K \text{ is affine } \forall K \in \mathcal{H}\}.$$

We denote by $(\psi_i)_{1\leq i\leq N}$ the basis function of V_h. We recall that, in this basis, $u_h \in V_h$ writes:

$$u_h(\mathbf{x}) = \sum_{i=1}^{n} v(\mathbf{p}_i)\, \psi_i(\mathbf{x}), \quad \forall \mathbf{x} \in \Omega_h.$$

Let K be a triangle and $[\mathbf{p}_0, \mathbf{p}_1, \mathbf{p}_2]$ its three vertices. If $\mathbf{x} \in K$, then we have:

$$u_h(\mathbf{x}) = u_h(\mathbf{p}_0)\, \psi_0(\mathbf{x}) + u_h(\mathbf{p}_1)\, \psi_1(\mathbf{x}) + u_h(\mathbf{p}_2)\, \psi_2(\mathbf{x}), \qquad [5.1]$$

where $\psi_i(\mathbf{x}) = \beta_i = \frac{A_{Ki}}{A_K}$ is the ith barycentric coordinates of $\mathbf{x}$ with respect to K.

5.3. Solving the problematic of the unknown solution (2D/3D)

For elliptic second-order problems, Céa's lemma (Ciarlet 1991) states that the finite element approximation error can be bounded by the interpolation error:

$$\|u - u_h\|_{H^1} \leq c\, \|u - \Pi_h u\|_{H^1},$$

where c a constant independent of the current mesh. Further, the right-hand side can be expressed in terms of the Hessian derivative of u. But this error is of first order with respect to mesh size. Also for elliptic problems, the Aubin–Nitsche analysis is written thus:

$$\|u - u_h\|_{L^2} \leq ch^2\, \|u\|_{H^2}.$$

In the multiscale analysis presented here, we assume that these relations still hold in the class of problems envisaged.

A second type of argument is the following. Given a numerical solution W_h, a solution of higher regularity $R_h(W_h)$ is *recovered*, so that the following interpolation error estimate (Chen et al. 2007; Loseille and Alauzet 2011b) holds (3D case):

$$\|R_h(W_h) - \Pi_h R_h(W_h)\|_{L^p} \leq N^{-\frac{2}{3}} \left(\int_\Omega \det\left(|H_{R_h(W_h)}|\right)^{\frac{p}{2p+3}} \right)^{\frac{2p+3}{3p}},$$

where $H_{R_h(W_h)}$ is the Hessian of the recovered solution and N an estimate of the desired number of nodes. Notation Π_h holds for the usual elementwise affine interpolation relying on values at vertices. When working in the continuous mesh framework (Loseille and Alauzet 2011a,b), we also have the optimal metric field leading to the previous error estimate:

$$\mathcal{M}_{\mathbf{L}^p} = \mathcal{K}_{L^1}(\left|H_{R_h(W_h)}\right|) = D_{\mathbf{L}^p}\left(\det\left|H_{R_h(W_h)}\right|\right)^{\frac{-1}{2p+3}}\left|H_{R_h(W_h)}\right|.$$

If anisotropic mesh prescription is naturally deduced in this context, interpolation-based methods do not take into account the features of the PDE. However, in some simplified context and assumptions (elliptic PDE, specific recovery operator R_h), we have

$$\|W - W_h\| \leq \frac{1}{1-\alpha}\|R_h(W_h) - \Pi_h R_h(W_h)\| \quad \text{with } 0 \leq \alpha < 1,$$

so that a good convergence to the exact solution may be observed (Loseille et al. 2007). Indeed, if $R_h(W_h)$ is a better approximate of W in the following sense:

$$\|W - W_h\| \leq \frac{1}{1-\alpha}\|R_h(W_h) - W_h\| \quad \text{with } 0 \leq \alpha < 1,$$

and if the reconstruction operator R_h has the property:

$$\Pi_h R_h(W_h) = W_h,$$

then we can bound the approximation error of the solution by the interpolation error of the reconstructed function $R_h(W_h)$:

$$\|W - W_h\| \leq \frac{1}{1-\alpha}\|R_h(W_h) - \Pi_h R_h(W_h)\|.$$

Note that from a practical point of view, $R_h(W_h)$ is never constructed, only its first and second derivatives are estimated. Standard recovery techniques include least-square, L^2-projection, the Green formula or the Zienkiewicz-Zhu recovery operator (Zienkiewicz and Zhu 1992). For all numerical examples, the $\mathbf{L}^2$ norm is used with the L^2-projection to recover derivatives. Actually, similar analyses based on the interpolation error show (practically) that the link between the interpolation error and the approximation error is even stronger, and we refer to Fortin (2000a). Last but not least, we shall show that, when extended to L^p analysis, the method offers a more powerful mesh adaptation tool. The multiscale criterion is then defined as

$$\mathcal{E}(\mathcal{M}) = \|u - \Pi_h u\|_{L^p} \ , \ 1 \le p < \infty.$$

Nevertheless, this criterion cannot be directly used since u is not known. The idea is to consider a higher order solution approximation u^* of u derived from u_h which is twice continuously differentiable. We do not assume that u^* is a better approximation of u than u_h but u^* can be built in such a way that its derivatives better approximates the derivatives of u. In other words, the interpolation error is approximated thus:

$$\|u - \Pi_h u\|_{L^p} \approx \|u^* - \Pi_h u^*\|_{L^p}.$$

Practically, only the Hessian of u^* is recovered, the purpose being to have

$$\|u - \Pi_h u\|_{L^p} \approx fcn(p, \mathcal{M}, H(u^*)).$$

In the context of discontinuous flows, the numerical solution, while approximating the discontinuous field, is still continuous piecewise linear by element. We consider that the mesh acts as a regularization operator on the solution and we approximate the solution u with a continuous higher order representation and apply our error estimate. The purposes of the following sections are, on the one hand, to obtain an approximation of the unavailable Hessian, and, on the other hand, to establish the estimate on the interpolation.

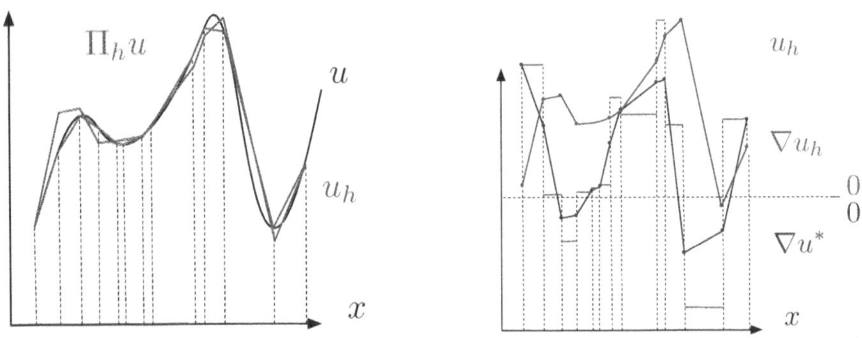

Figure 5.2. *Left: Illustration of exact solution u (black), its linear interpolate $\Pi_h u$ (red) and numerical solution u_h (green). Right: Illustration of piecewise constant gradient ∇u_h (blue) and piecewise linear recovered gradient ∇u^* (purple). For a color version of this figure, see www.iste.co.uk/dervieux/meshadaptation1*

5.4. Numerical computation/recovery of the Hessian matrix

One key point in the mesh adaptation scheme is related to the construction of the metric tensor and, more precisely, to the evaluation of the Hessian matrix. Indeed, constructing a good (reasonable) metric requires computing the Hessian matrix accurately. Various techniques have been investigated so far. In the following, we present three of them.

5.4.1. *Numerical computation of nodal gradients (2D)*

Let K be a triangle and $[\mathbf{p}_0, \mathbf{p}_1, \mathbf{p}_2]$ its three vertices. If $\mathbf{x} \in K$, relation [5.1] provides the values $u_h(\mathbf{x})$. Similarly, the gradients $\nabla u_h(\mathbf{x})$ is given by

$$\nabla u_h(\mathbf{x}) = u_h(\mathbf{p}_0)\,\nabla\psi_0(\mathbf{x}) + u_h(\mathbf{p}_1)\,\nabla\psi_1(\mathbf{x}) + u_h(\mathbf{p}_2)\,\nabla\psi_2(\mathbf{x}), \qquad [5.2]$$

where

$$\begin{aligned}
\frac{\partial\psi_0}{\partial x} &= \frac{y_1 - y_2}{2\,A_K} & \qquad \frac{\partial\psi_0}{\partial y} &= \frac{x_2 - x_1}{2\,A_K}\\
\frac{\partial\psi_1}{\partial x} &= \frac{y_2 - y_0}{2\,A_K} & \qquad \frac{\partial\psi_1}{\partial y} &= \frac{x_0 - x_2}{2\,A_K}\\
\frac{\partial\psi_2}{\partial x} &= \frac{y_0 - y_1}{2\,A_K} & \qquad \frac{\partial\psi_2}{\partial y} &= \frac{x_1 - x_0}{2\,A_K}.
\end{aligned}$$

We deduce that ∇u_h is piecewise constant, that is, constant on each element K. We denote by $\nabla u_{h|K}$ the constant gradient on element K.

However, ∇u_h is not defined at mesh vertices. We propose to recover nodal gradients for the piecewise constant representation of ∇u_h using a $\mathbf{L}^2$-projection method. The local $\mathbf{L}^2$-projection operator is based on the Clément interpolation operator (Clément 1975). Let $\mathbf{p_i}$ be a vertex of mesh $\mathcal{H}$. We denote by S_i the support of basis function ψ_i:

$$S_i = supp\,\psi_i.$$

We introduce the approximation spaces of the piecewise constant and linear function:

$$\begin{aligned}
V_h^0 &= \{v \in \mathbf{L}^2(\Omega) \mid v_{|K} \in \mathbb{P}^0 \;\; \forall K \in \mathcal{H}\}\\
V_h^1 &= \{v \in \mathcal{C}^0(\Omega) \mid v_{|K} \in \mathbb{P}^1 \;\; \forall K \in \mathcal{H}\},
\end{aligned}$$

where $\mathbb{P}^0$ and $\mathbb{P}^1$ are the set of constant and linear polynomials. The idea is to find, in a $\mathbf{L}^2$-norm sense, the best constant gradient $\nabla_R u_h$ approximating the piecewise constant field ∇u_h on S_i. More precisely, for $v \in \mathbf{L}^2(\Omega)$, we define $\Pi_0 v \in V_h^0$ thus:

$$\forall S_i \subset \mathcal{H}, \quad \begin{cases} (\Pi_0 v)_{|S_i} \in \mathbb{P}^0 \\ \displaystyle\int_{S_i} (\Pi_0 v - v)\; w = 0, \ \forall w \in \mathbb{P}^0. \end{cases}$$

DEFINITION 5.1.– The Clément interpolation operator $\Pi_c : V_h^0 \longrightarrow V_h^1$ is defined thus Clément (1975):

$$\Pi_c v := \sum_{i=1}^{n} \Pi_0 v(\mathbf{p}_i)\, \psi_i.$$

Using Clément interpolation operator, we now describe how we recover nodal gradients from $\nabla u_h \in \mathbb{P}^0$. For each $S_i \subset \mathcal{H}$, we have for the particular choice $1 \in \mathbb{P}^0$:

$$\begin{aligned} \int_{S_i} (\Pi_0(\nabla u_h) - \nabla u_h) = 0 \iff & \int_{S_i} \Pi_0(\nabla u_h) = \int_{S_i} \nabla u_h \\ \iff & |S_i|\, \Pi_0(\nabla u_h)_{|S_i} = \sum_{K_j \in S_i} \int_{K_j} \nabla u_h \\ \iff & \Pi_0(\nabla u_h)_{|S_i} = \frac{\sum_{K_j \in S_i} |K_j|\, \nabla u_{h|K_j}}{|S_i|}, \end{aligned}$$

where $|K|$ and $|S_i|$ denote the volume of element K and support S_i, respectively. For each vertex $\mathbf{p}_i$, we thus have the following gradient reconstruction:

$$\nabla_R\, u_h(\mathbf{p}_i) = \frac{\sum_{K_j \in S_i} |K_j|\, \nabla u_{h|K_j}}{|S_i|}.$$

In fact, this procedure is equivalent to a reconstruction by means of a volume-weighted averaging. The recovery procedure provides us with gradient nodal values and thus we get a piecewise linear representation of the gradient because of the Clément interpolation operator. Similar relations are derived in three dimensions.

5.4.2. *A double* L^2*-projection method*

We can use the $\mathbf{L}^2$-projection method to recover the Hessian of u^* from u_h. To this end, the recovery described above is simply applied to each component of the gradient. A more detailed formulation along with some comparisons to other methods is available in Alauzet and Loseille (2010a).

5.4.3. *A method based on the Green formula*

The Hessian of the solution can be recovered using a weak formulation based on the Green formula. We consider the same notations as previously. For each vertex $\mathbf{p}_k$ of $\mathcal{H}$, we have for $1 \leq i, j \leq 3$:

$$\int_{\mathcal{H}} \frac{\partial^2 u_h}{\partial x_i \partial x_j} \varphi_k = \int_{S_k} \frac{\partial^2 u_h}{\partial x_i \partial x_j} \varphi_k = -\int_{S_k} \frac{\partial u_h}{\partial x_j} \frac{\partial \varphi_k}{\partial x_i} + \int_{\partial S_k} \frac{\partial u_h}{\partial n} \varphi_k \, d\sigma$$
$$= -\sum_{K \in S_k} \int_K \frac{\partial u_h}{\partial x_j} \frac{\partial \varphi_k}{\partial x_i},$$

as the shape function is zero on the boundary of the stencil ∂S_k. Each component of the Hessian is then recovered with the relation:

$$\frac{\partial^2 u^*}{\partial x_i \partial x_j}(\mathbf{p}_k) := \frac{-\displaystyle\int_{S_k} \frac{\partial u_h}{\partial x_j} \frac{\partial \varphi_k}{\partial x_i}}{\displaystyle\int_{S_k} \varphi_k} = -\frac{\displaystyle\sum_{K \in S_k} (\frac{\partial u_h}{\partial x_j})|_K \int_K \frac{\partial \varphi_k}{\partial x_i}}{\frac{|S_k|}{4}},$$

which is equivalent to lump the mass matrix of the left-hand side of the previous relation.

5.4.4. *A least-square approach*

Let $\mathbf{p}$ be a mesh vertex and let u be the solution taken into account to define the metric. By considering a Taylor expansion of u at a vertex $\mathbf{p}_i$ connected to the vertex $\mathbf{p}$ (i.e. $\mathbf{p}_i \in \mathcal{B}(\mathbf{p})$, the ball of $\mathbf{p}$) and truncated at the order 2, we can write the following relation:

$$u_i = u + \mathbf{pp}_i . \nabla u(\mathbf{p}) + \frac{1}{2} \langle {}^t\mathbf{pp}_i, H_u(\mathbf{p})\, \mathbf{pp}_i \rangle \;\Leftrightarrow\; \frac{1}{2} \langle {}^t\mathbf{pp}_i, H_u(\mathbf{p})\, \mathbf{pp}_i \rangle$$
$$= u_i - u - \mathbf{pp}_i . \nabla u(\mathbf{p})$$

with the notations $u = u(\mathbf{p})$ and $u_i = u(\mathbf{p}_i)$. This relation can be developed using the notations

$$\mathbf{pp}_i = {}^t\!(x_i\ y_i\ z_i)\,, \qquad \nabla u(\mathbf{p}) = {}^t\!(\alpha\ \beta\ \gamma)\,, \qquad H_u(\mathbf{p}) = \begin{pmatrix} a & b & c \\ b & d & e \\ c & e & f \end{pmatrix}$$

as follows:

$$\frac{1}{2}(ax_i^2 + 2bx_iy_i + 2cx_iz_i + dy_i^2 + 2ey_iz_i + fz_i^2) = u_i - u - (\alpha x_i + \beta y_i + \gamma z_i). \tag{5.3}$$

This leads to a usually over-determined system[1] of the form

$$AX = B, \quad \text{with } {}^tX = (\,a\ b\ c\ d\ e\ f\,),$$

where A is a $n \times 6$ matrix ($n = Card(\mathcal{B}(\mathbf{p}))$) function of (x_i, y_i, z_i) and B is a vector of dimension n given by the right-hand side of the relation [5.3], and function of $(\alpha, \beta, \gamma, x_i, y_i, z_i, u, u_i)$. This system is solved using a least-square approximation, that is, it consists of minimizing the distance between the vectors AX and B of $\mathbb{R}^6$ by minimizing the square of the Euclidean norm of their difference. The problem is thus:

$$\text{Find } u \in \mathbb{R}^6 \text{ such that } \|AX - B\|^2 = \inf_{Y \in \mathbb{R}^6} \|AY - B\|^2.$$

It can be shown that the solution of this problem is the solution of the linear 6×6 system of normal equations:

$${}^tAAX = {}^tAB.$$

The latter is then solved using a standard Gauss method.

REMARK 5.1.– If, in some peculiar cases, the system is under-determined (i.e. $Card(\mathcal{B}(\mathbf{p})) < 6$), additional vertices connected to the vertices of $\mathcal{B}(\mathbf{p})$ can be taken into account.

1 The system is overdetermined as six coefficients must be computed and the vertex $\mathbf{p}$ is usually connected to more than six vertices $\mathbf{p}_i$ in three dimensions.

5.4.5. *From our experience*

The behavior of the Green formula on shocks appeared too delicate. Least squares are accurate for isotropic meshes but not sufficiently robust for strongly anisotropic meshes. The double L^2-projection is both robust and accurate.

5.4.6. *Discrete-continuous interpolation*

While the interpolation of a continuous function (corollary 4.1) needs the knowledge of the Hessian in any point of the computational domain, the knowledge of a discrete field u_h and the evaluation of the Hessian on every vertex allow to compute what we call the *discrete-continuous interpolation* $\pi_h u_h$, defined as follows:

– the discrete-continuous interpolation error $\pi_h u_h - u_h$ is a P_2 polynomial on each element;

– it is vanishing on vertices;

– on any mid-edge ij, we have

$$(\pi_h u_h - u_h)(\mathbf{x_i} + \mathbf{x_j}) = -\frac{1}{2}(H_h(u_h)(\mathbf{x}_i) + H_h(u_h)(\mathbf{x}_j))(\mathbf{x_i} - \mathbf{x_j})^{2/4}.$$

If the Hessien is uniform, this discrete-continuous interpolation is identical to the continuous interpolation of corollary 4.1.

REMARK 5.2.– This definition is coherent with the one used in Vassilevski and Agouzal (2005). □

5.5. Solution interpolation

5.5.1. *Localization algorithm*

The localization problem or research of point location consists of identifying the element of a simplicial mesh containing a given point. The localization of a given point in a mesh is a frequent issue that arises in various situations. With regard to interpolation methods, we initially have a mesh with a field, here the solution, that we call *background mesh*, denoted $\mathcal{H}^{back}$. We aim at transferring or interpolating the field onto another mesh called *current mesh* or *new mesh*, denoted $\mathcal{H}^{new}$. Therefore, the algorithm consists of finding which elements of the background mesh contain the vertices of the new mesh in order to apply an interpolation scheme.

Here, we consider the simplified problem where the background and the new meshes are discretizations of the same domain Ω. Otherwise, even the case of simplicial meshes lead to handle difficult configurations. Indeed, background and current meshes can be non-convex and can contain holes. It is also possible that the overlapping of the current mesh does not coincide with the background mesh since their boundary discretization can differ. Consequently, some vertices of the current mesh can be outside of the background mesh and conversely. Moreover, efficient localization algorithms have to be implemented to avoid the naive quadratic algorithm in $O(N_{ver}^{new} \times N_{tri}^{back})$ where N_{ver}^{new} is the number of vertices of $\mathcal{H}^{new}$ and N_{tri}^{back} the number of triangles of $\mathcal{H}^{back}$.

The localization can be solved efficiently by traveling across the background mesh using its topology, that is, the neighboring elements of each element, because of a barycentric coordinates-based algorithm (Frey and George 2008; Löhner 2001). More precisely, in two dimensions, let $\mathbf{p}$ be a vertex of the new mesh, $K = [\mathbf{p}_0, \mathbf{p}_1, \mathbf{p}_2]$ a triangle of the background mesh. From the signs of the three barycentric coordinates $\{\beta_i\}_{i=0,2}$, three possible cases arise (see Figure 5.3):

– if all barycentric coordinates are positive, then vertex $\mathbf{p}$ is located inside element K;

– if one barycentric coordinate is negative, then it indicates the direction for the next move. For instance, if barycentric β_i is negative, then we move to neighboring element K_i sharing edge $\mathbf{e}_i$ with K. We say that $\mathbf{p}$ is viewed by edge $\mathbf{e}_i$;

– if two barycentric coordinates are negative, then two neighboring triangles are possible for the next move. A random choice or a geometric one is used.

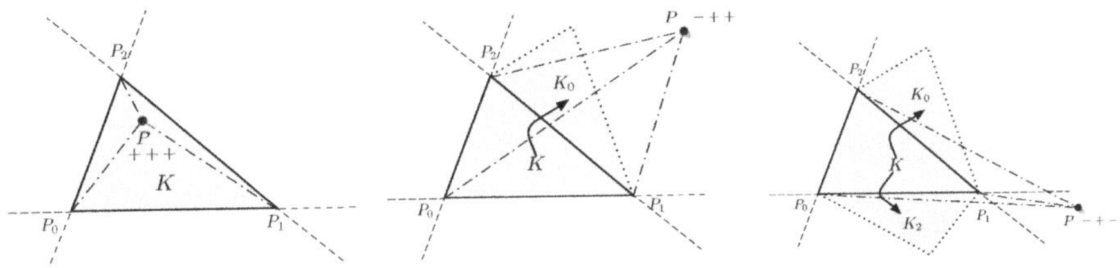

Figure 5.3. *Illustration of the three possible cases depending on the signs of the three barycentric coordinates of vertex* $\mathbf{p}$ *with respect to triangle* K *when moving inside the background mesh*

Starting from an initial element K_0 of the background mesh, we apply the previous test. According to the signs of the barycentric coordinates, we pass through the corresponding neighbor of K_0 and we repeat this process until the three barycentric coordinates are positive meaning that the visited triangle contains $\mathbf{p}$. With this algorithm, we follow a path in the background mesh to locate vertex $\mathbf{p}$ as shown in Figure 5.4 (left). This algorithm complexity is in $O(n \times N_{ver}^{new})$ where n is the average number of visited triangles for each path. However, cyclic or closed paths can occur. The element containing the vertex is missed and an infinite loop is obtained. In this case, the path leads us to an already tested element, as presented in Figure 5.4 (right). In this academic example, starting from K_0, triangles K_1, K_2, K_3, K_4 and K_5 are visited bringing us back to K_0. A color algorithm, to mark already visited elements, is used to avoid this problem allowing us to choose another direction when several choices reoccur. Another way to solve this problem is to consider a random choice when several possibilities occur.

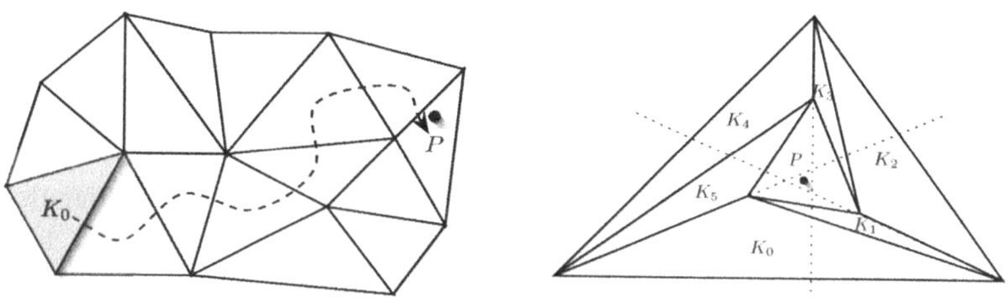

Figure 5.4. *Left: A possible path to locate the vertex* $\mathbf{p}$ *of the new mesh starting from the triangle* K_0 *of the background mesh. Right: Cyclic path leading us to an already checked element. Starting from* K_0*, triangles* K_1*,* K_2*,* K_3*,* K_4 *and* K_5 *are visited, bringing us back to* K_0

Another difficulty arises when the path is stuck by the geometry of domain Ω. Starting element K_0 and vertex $\mathbf{p}$ are separated by a hole or by a non-convex domain (see Figure 5.5, left). The path demands to pass through the hole or the boundary to reach the element containing vertex $\mathbf{p}$. A simple, but inefficient, way to remedy this problem is to make an exhaustive search, that is, for such a vertex all elements of the background mesh are tested. Besides, a more challenging solution is to follow a path on the boundary in order to bypass the obstacle.

5.5.1.1. *Localization coupled with a grid structure*

The previous algorithm can be very time consuming if a large number of elements (e.g. n is large) needs to be visited between triangle K_0 and the solution triangle. This can result in a large number of area computations. This major drawback leads us to consider a more local approach combining the algorithm with a grid structure (a tree-like structure can be considered). This facilitates and speeds up the localization process. A grid enclosing the mesh is constructed and, for each grid cell, one element of the background mesh located in it, if any, is recorded. Then, to locate a new vertex in the background mesh, the cell containing the vertex is first identified and then the localization scheme starts from the element associated with this cell. In this way, the number of visited triangles is reduced and the number of necessary computations decreases as well. In the case where we are stuck by the boundary, because of a hole or a non-convex domain, the grid structure helps us to bypass the obstacle. Indeed, elements associated with grid neighboring cells of the current one are considered as new initial guesses for the searching algorithm. The localization is restarted from one of these new elements.

REMARK 5.3.– Note that the grid (or the tree-like structure) could be defined in various ways depending on the nature of the data set. In this respect, for a grid, the number of cells and thus the occupation of the cells are parameters that clearly affect the efficiency of the whole process.

5.5.1.2. *Localization using the topology of both meshes*

We can even improve the locality of the localization scheme by using the topology of both meshes. The previous algorithm can be very time consuming if a large number of elements needs to be visited between triangle K_0 and the solution triangle. This major drawback leads us to consider a more local approach by using the topology of both meshes. Such scheme tends to minimize n the number of elements visited when locating new vertices. Instead of determining the location of the new vertices in their data (or storage) order, the idea is that once a vertex $\mathbf{p}$ has been located in a background element K, then we handle the set of vertices $\{\mathbf{q}_i\}_{i=1,m}$ of the ball of $\mathbf{p}$, that is, the set of vertices that are connected to $\mathbf{p}$ by an edge. For the vertices $\{\mathbf{q}_i\}_{i=1,m}$, we set as starting element of the localization process the triangle K that contains $\mathbf{p}$ (see Figure 5.5, right). Consequently, the number of visited triangles is drastically reduced as in this algorithm the initial guess of the searching process is at the element (or connectivity) level. Moreover, with this approach, the scheme does not depend on any parameters. Another advantage of this approach is that this scheme avoids the problem where the process is stuck by a hole or a non-convex boundary as vertices $\{\mathbf{q}_i\}_{i=1,m}$ are connected to $\mathbf{p}$ in the new mesh. This algorithm is also in $O(n \times N_{ver}^{new})$ where n is the average number of visited triangles and here n tends to be optimal. Indeed, in practice the number of visited triangles is on average less than 3. In fact, n is of the order of the number of elements of the background mesh that are overlapped by an element of the current mesh.

Two problems still have to be handled with care: the "fork" and the boundary problems. The first occurs when the path reaches a geometrical fork or a crossroads with multiple choices for the next move, then the presented algorithm could make the wrong choice and ask to process in the wrong direction. The second problem is that this algorithm is not able to handle the case where vertices are located outside of the background discretized domain. Specific ad hoc treatments can be considered to solve such issues. At worst, it is always possible to perform an exhaustive search.

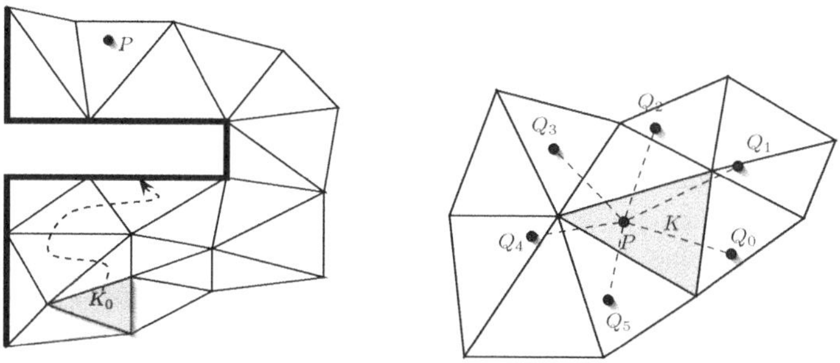

Figure 5.5. *Left: Starting element K_0 and vertex $\mathbf{p}$ are separated by the non-convex domain. The path demands to pass through the boundary to reach the element containing vertex $\mathbf{p}$. Right: The number of visited triangles in the localization scheme is reduced by using the topology of both meshes. Vertex $\mathbf{p}$ has been located in element K. Then, the set of vertices $\{\mathbf{q}_i\}_{i=1,m}$ connected to $\mathbf{p}$ uses element K as initial guess for the localization scheme*

5.5.2. *Classical polynomial interpolation*

In this section, we present several interpolation algorithms, which are not conservative in the sense that the integral of interpolated functions do not coincide with the integral of the primitive function. The provided solution is considered to be piecewise linear by element and continuous. Let us denote by $\mathbf{p}$ a vertex of the current mesh, $K = [\mathbf{p}_0, \mathbf{p}_1, \mathbf{p}_2]$ a triangle of the background mesh containing $\mathbf{p}$ and β_i, for $i = 0, ..., 2$, the barycentric coordinates of $\mathbf{p}$ with respect to K. In the following, we denote

– $\mathbb{P}^k$ to be the set of polynomials of degree less or equal than k;

– $\mathbb{P}_r$ to be the set of polynomials where the solution given by the interpolation scheme lies.

5.5.2.1. *Linear interpolation*

The easiest interpolation operator is the classical $\mathbb{P}^1$ interpolation, and in terms of barycentric coefficients β_i, it is written thus:

$$u(\mathbf{p}) = \sum_{i=0}^{3} \beta_i(\mathbf{p})\, u(\mathbf{p}_i).$$

This operator gives exactly u when u is $\mathbb{P}^1$, and it is of order 2. This operator does not conserve the mass. Indeed, if an edge between two triangles is *swapped* then this interpolation keeps unchanged solution at the triangles vertices, whereas the mass of the solution has changed.

5.5.2.2. *Quadratic interpolation*

The proposed quadratic scheme uses $\mathbb{P}^2$ Lagrange test functions in triangle K to reconstruct a quadratic representation of the solution on K. This interpolation requires the solution nodal value at triangle vertices $\mathbf{p}_0$, $\mathbf{p}_1$ and $\mathbf{p}_2$, and the solution at the triangle mid-edges. We denote by $\mathbf{p}_3$, $\mathbf{p}_4$ and $\mathbf{p}_5$ the middle of edges $\mathbf{e}_2$, $\mathbf{e}_0$ and $\mathbf{e}_1$, respectively (i.e. $\mathbf{p}_i$ is the middle of edge $\mathbf{p}_{[i]}\mathbf{p}_{[i+1]}$, for $i = 3, ..., 5$). The quadratic scheme is given by

$$u(\mathbf{p}) = \sum_{i=0}^{5} \psi_i(\mathbf{p})\, u(\mathbf{p}_i) \text{ with } \begin{cases} \psi_i(\mathbf{p}) = \beta_i(\mathbf{p})\,(2\,\beta_i(\mathbf{p}) - 1) & \text{for } i = 0, ..., 2, \\ \psi_i(\mathbf{p}) = 4\,\beta_{[i]}(\mathbf{p})\,\beta_{[i+1]}(\mathbf{p}) & \text{for } i = 3, ..., 5. \end{cases}$$

If the mid-edge values are known, then this interpolation is $\mathbb{P}^2$ exact and hence of order 3. However, in our case the solution representation is continuous and piecewise linear by element. Therefore, we have to specify how the mid-edge solution values are obtained. From the solution nodal values, gradients at vertices could be reconstructed (see previous section). Let us denote by $\nabla u(\mathbf{p}_i)$ the gradient for vertex $\mathbf{p}_i$. In this case, we have an over-determined system for the quadratic reconstruction. Consequently, we choose to perform a cubic reconstruction on each edge to get the mid-edge values using the solution and the gradient at each extremity. After some algebra, mid-edge values are given thus:

$$u(\mathbf{p}_i) = \frac{u(\mathbf{p}_{[i]}) + u(\mathbf{p}_{[i+1]})}{2} + \frac{\nabla u(\mathbf{p}_{[i]}) - \nabla u(\mathbf{p}_{[i+1]})}{8} \cdot \mathbf{p}_{[i]}\mathbf{p}_{[i+1]}, \text{ for } i = 3, ..., 5.$$

The order and the exactness of the scheme depends on the gradient reconstruction. Nonetheless, for the proposed approach we have $\mathbb{P}^1 \subset \mathbb{P}_r \subset \mathbb{P}^2$ and an order between 2 and 3 is obtained. The reconstruction is $\mathcal{C}^1$ except through the elements edges. This interpolation is not conservative. The problem of the above *swap* example holds. Moreover, this reconstruction does not verify the maximum principle. Hence, we have to limit this reconstruction in order to avoid the creation of new extrema. The proposed limiter is simply

$$u_{lim}(\mathbf{p}) = \max(\min(u(\mathbf{p}), u_{max}), u_{min}), \text{ with } u_{min} = \min_{i=0..2} u(\mathbf{p}_i)$$

$$\text{and } u_{max} = \max_{i=0..2} u(\mathbf{p}_i).$$

REMARK 5.4.– The quality of transfers between meshes can be important for the steady adaptation loop and crucial for the unsteady adaptation loop. Let us stress that *conservative transfers* are proposed in Alauzet and Mehrenberger (2010b). □

5.6. Mesh adaptation algorithm

Mesh adaptation is a nonlinear problem linked to the solution. We have to propose an iterative procedure to solve this problem. For steady simulations, an adaptive computation is carried out via a mesh adaptation loop inside which an algorithmic (or iterative) convergence of the couple mesh-solution is sought, in the sense that the solution is algorithmically (or iteratively) converging toward the steady-state solution and the mesh is converging toward the adapted mesh associated to this converged steady state. In a *feature-based approach*, the user chooses a particular solution-dependent field, the *sensor* (or feature). If the solution field is a scalar field, the solution is taken as sensor. This is sketched in algorithm 5.1.

Algorithm 5.1. Feature-based mesh adaptation for PDE. This loop provides a couple $(mesh, flow\ solution)$ for a prescribed number of nodes N according to the Hessian of a feature, the sensor, prescribed by user.

Initial mesh $\mathcal{H}_0$, solution u_0, and complexity N.
while i $\leq n_{adap}$ **do**

1) Compute optimal metric for the sensor interpolation error estimate and complexity $\Longrightarrow \mathcal{M}_i$

2) Generate new adapted mesh $\Longrightarrow \mathcal{H}_i$

3) Interpolate state on the new mesh $\Longrightarrow (u^0)_i$

4) Compute state $\Longrightarrow u_i$

5) **if** (solution not fully adapted) **then** $i = i + 1$ **else Stop**

At each stage, a numerical solution is computed on the current mesh and is analyzed by means of the error estimate. In our case, the metric is obtained via the geometric error estimate using the recovery. Next, a unit mesh is generated with respect to this metric. Finally, the solution is interpolated on the new adapted mesh. This procedure is repeated until the algorithmic convergence of the solution and the mesh is achieved.

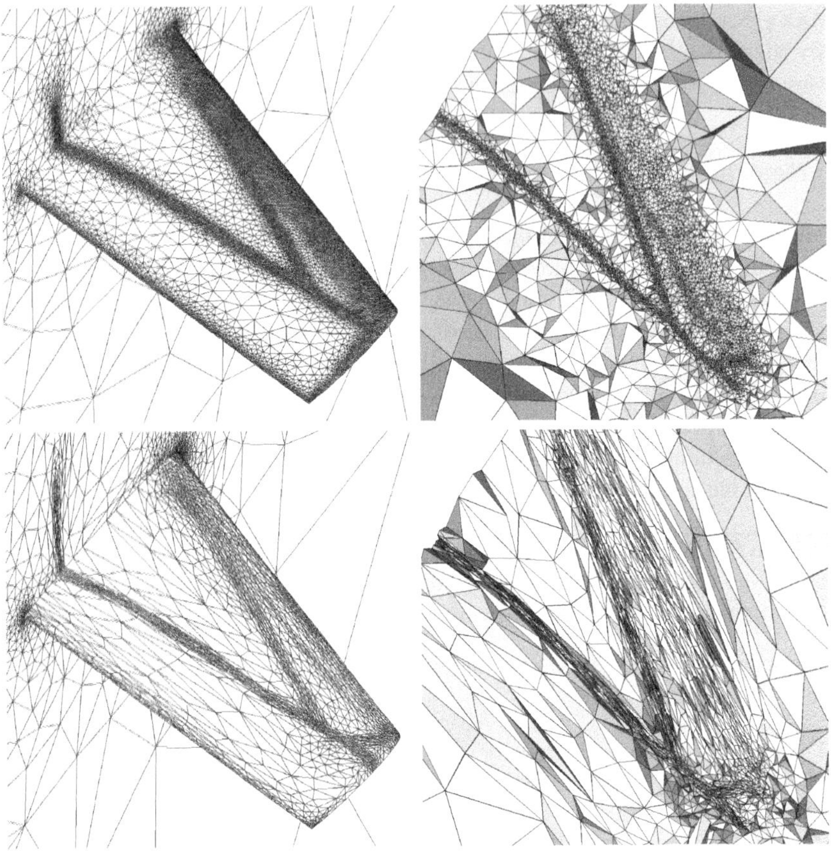

Figure 5.6. *Onera M6 wing test case. Top: Isotropic mesh adaptation and, bottom, anisotropic mesh adaptation. Left: Adapted surface mesh and, right, cut through the volume adapted mesh*

5.7. Example of a CFD numerical simulation

This CFD example concerns a classic numerical simulation of transonic air flow around the ONERA M6 wing. We assume that the flow is inviscid, it is thus modeled by the compressible Euler equations. A second-order finite volume solver (Code Wolf) is used to compute the numerical solution (see Chapter 1). The flow is computed for Mach number equal to 0.8395 with an angle of attack of 3.06 degrees. This transonic simulation case results in a well-known lambda-shock.

The mesh is adapted using algorithm 5.1. with the L^2 norm of the Mach number as error to minimize. The surface and volume meshes have been adapted using `Yams` (Frey 2001) and `Gamanic3d` (George 1999). The mesh has been adapted nine times, every 250 time steps. Figure 5.6 shows the adaptation in the isotropic and anisotropic cases. The final isotropic mesh contains 231,113 vertices and 1,316,631 tetrahedra and the final anisotropic mesh contains 23,516 and 132,676 tetrahedra. In this example, the maximal aspect ratio achieved for the anisotropic elements is about 10. Nevertheless, the anisotropic metric leads to a dramatic reduction of the number of degrees of freedom, roughly one order less than in the isotropic case (for the same error level).

5.8. Conclusion

This chapter has described the basic multiscale mesh adaptation algorithm. It has described the different tools for running this algorithm, and mainly, the recovery methods producing the local error, and the re-interpolations between a mesh and the new mesh.

5.9. Notes

5.9.1. *A short review of mesh/PDE coupling*

The coupling between a discrete PDE and its adapted mesh is nonlinear and two-way. The linearity of the influence of the mesh on the discrete PDE solution is definitively not true (finite element matrix depend nonlinearly on vertices coordinates). Further, the mesh is generally not defined in a vector space, it can be deformed, but also enriched. Then, it is useful to examine the way this coupling can be taken into account. Starting from a given mesh, we can compute a PDE solution. We then derive information from the discrete PDE solution in order to change the mesh. Let us review here the way the mesh is changed. Here, we call a mesh a rather complete description of the way the PDE is discretized and we do not discuss unsteady issues (Chapter 2 of Volume 2).

– Insufficient accuracy can be interpreted as a too large spatial step (FDM) or element size (FEM) in some region of the computational domain. This can be improved by

a) moving mesh nodes in order to concentrate them on the unsufficient accuracy region, and we call this r-adaptation;

b) adding nodes or elements in the considered region, and we call this the h-adaptation;

c) we can try to cure the local unsufficient accuracy by locally increasing the order p of the approximation; this is the basis of p-adaptation.

The r-adaptation methods. The r-adaptation (r probably holds for relocation) does not change the number of nodes and their connectivity, but only their coordinates. In order to do this adaptive mesh deformation, it is necessary to propose a field of displacements of the nodes. A central problem is then how to deform without losing the admissibility of mesh. By admissible we mean that the mesh is valid for the considered approximation: with no mesh overlapping, typically. This means that the deformation map is a regular one-to-one map. In some case, the approximation error criterion chosen for mesh deformation is able to maintain the mesh admissibility, or does it after minor changes in the method. This is the case for the moving finite element, in which the variational principle of finite elements is extended to the definition of optimal node location. We refer to the monograph by Baines (1994). In most case, the approximation error criterion is not able to maintain the mesh admissibility. It is necessary to combine it with a mesh model reinforcing the bijective character of the underlying mapping. From this standpoint, r-adaptation is connected with the important topic of mesh deformation techniques applied to moving geometries, and used in conjunction with Arbitrary–Lagrange–Euler formulations. Mesh deformation techniques are generally inspired by elasticity deformation models, either by discretizing a physical principle or by building a discrete system from a physical analogy, which relies on mesh features such as edges. For the second option, let us mention some typical references: The attraction spring system of Gnoffo (1983), the expansive spring system of Batina (1989) and the torsional spring system of Farhat et al. (1998). In Gnoffo (1983), a directional error sensor is evaluated on each edge of the mesh. It is introduced as a spring strength in the mesh system. A relaxation solves the mesh system with some coupling with the discrete PDE. This kind of approach can produce mesh stretching and is very efficient for the capture of thin gradient layers in a flow and was applied for example to re-entry flow prediction. The limitation due to a constant topology is in most case delicate to compensate by introducing some phases of topology change in the deformation algorithm. After the mesh is moved, an adequate transfer of the PDE solution from previous mesh to new one is useful for steady prediction.

The h-adaptation methods. In h-adaptation, the mesh is locally replaced by a finer embedded one, by dividing regularly the initial mesh. Then the calculation may restart after a projection of the previous solution, for saving iterative convergence, according to usual embedded refinement principle. The point is to consider on which support the new solution is computed. The PDE approximation *a priori* applies only on conformal meshes. Here, "conformal" refers to the obtention to an admissible mesh for the standard version of the approximation. For example, a finite element conformal mesh in triangles is a mesh in which the domain is partitioned in triangles, with intersection between two triangle closures are empty, a vertex, an edge of a triangle. In conformal h-adaptation (typically with finite elements), a new conformal mesh is built by re-building a conformal connectivity at the interface between the previous mesh and the piece of mesh resulting from the uniform division applied to the to-be-refined region. In a second family of methods, we get a new mesh, of extended type, the hanging nodes at the interface being accounted by the approximation in an ad hoc way. In general, this is done in the structure of a multi-level mesh (Berger and Colella 1989; Quirk 1991). It can be also done in a Chimera formulation (Steger et al. 1983). A last way is meshless methodology in which the dicretization primary support is a cloud of nodes, the connectivity being re-computed each time it is necessary (Morinishi 2001).

The p-adaptation methods. p-adaptation consists of varying the approximation order without changing the underlying geometrical mesh. However, the local number of degree of freedom is generally changed. This idea is generally combined, in an h–p-adaptation, with an h-adaptation principle (Oden 1989).

6

Multiscale Convergence and Certification in CFD

While the previous chapters explain how the mesh adaptation loop replaces a mesh by a new mesh, with the same complexity/number of nodes, but optimal in some sense, this chapter studies the search of a sufficiently mesh-converged computation. The mesh-convergence algorithm is an outer loop containing the mesh adaptation algorithm. As an example, convergence is analyzed for an academic case and the new algorithm is demonstrated with the flow around a supersonic aircraft shape.

6.1. Introduction

The previous chapter proposed an algorithm for replacing a given mesh by an adapted mesh with a prescribed complexity. The accuracy is better than with a non-adaptive mesh but the user cannot evaluate if the result is of sufficient accuracy or not. The purpose of this chapter is to use the previous chapters outputs in order to build a mesh adaptation method for PDEs producing certifiable results. Results are certified via mesh convergence measurements. Before doing this, we would like to examine what are the needs for an increased accuracy, in particular in compressible CFD models. While numerical methods and computers show impressive improvements, the numerical quality of CFD computations remains a delicate problem (see, for example, Hutton and Casey 2001). In particular, estimating approximation errors is one of today's most important challenges in CFD (Rumsey et al. 2018; Morrison 2022). Existing error estimates are progessively more useful for answering this question (see Becker et al. 1999; Fortin 2000b; Giles and Suli 2002), but useable estimates are demanding a certain degree of mesh convergence. Another approach is to directly measure mesh convergence and is still the most popular tool. Mesh convergence theory and methods date back at least to Romberg (1955). They have been well studied, but in compressible CFD, mesh convergence studies are still very computer intensive.

The fundamental assumptions of the standard mesh convergence theory are

1) uniform order of convergence;

2) structured Cartesian or body fitted;

3) embedded meshes.

The uniform order assumption is, for example, the basis of the Grid Convergence Index proposed in Roache (1994). Now a particular property of compressible flows is that they involve either strong singularities, such as shocks, or, in the viscous case, strong local gradients (viscous shocks, boundary layers). The uniform second-order assumption does not apply to the computation of these heterogeneities. Indeed, shock capturing schemes produce locally varying convergence orders. In the case of viscous flows, numerical order can be even more intricate since on very fine meshes, the numerical convergence will finally switch from first order to second order in shock region. But global second-order mesh convergence seems to be unattainable in most practical applications. Let us analyze in greater detail the two contexts we have identified.

For the *case of genuine discontinuities*, thousands of papers were devoted during decades to the design of quasi second-order accurate TVD schemes for the Euler model that are able to compute shocks stably with high resolution. However, even a high resolution scheme is unable to converge at second order toward a discontinuous solution when it is combined with a uniform mesh refinement process (see, for example, Carpenter and Casper (1999)). This failure is related to the impossibility in interpolating accurately a discontinuous solution on a sequence of uniformly divided meshes. We provided evidence in Chapter 2 regarding the poor performance of P_1 interpolation of discontinuous functions in uniform mesh refinement. Convergence order in L^2 norm is less than one. It is shown in that chapter that, conversely, *adequate sequences of adapted meshes* can offer a higher order convergence. But, in two or three dimensions, only the anisotropic mesh adaptation has the potential to produce a second-order convergence. We analyzed, for the interpolation problem in Chapter 4, why and in which conditions anisotropic adaptation are able to produce second-order accuracy on discontinuities.

In contrast, in the *case of viscous shocks*, it is true that second-order convergence should be always obtained when uniform refinement involves sufficiently fine meshes. But, in practical cases, probably because meshes are not sufficiently fine, we generally do not observe the second-order convergence. This illustrates that the usual asymptotical theory does not explain today's experience and needs to be extended. Our standpoint relies on the following remarks: the "viscous heterogenities" should be understood as behaving in some case as pure discontinuities, and in some case as smooth regions. Indeed, when the mesh which is used is not sufficiently fine to capture the smooth thickness of the shock, then the viscous shocked flow calculations generally show a convergence order, which is not better than the inviscid one, that is with a numerical order close to one. The key goal of this work is then to identify an algorithm that will show *early second-order convergence* of viscous flows. And the key principle is to use for viscous flows the same mesh adaptive method that is able to produce second-order accuracy on genuine discontinuities.

The main object of this chapter is to propose a combination of these methods that first will produce adapted solutions for a *fixed number of nodes*, and second will lead to *early second-order mesh convergence* when increasing this number. This will allow for a Romberg/Richardson-type numerical quality analysis: once the second-order numerical order is *certified* we derive classically an estimate of the approximation error. In the next section, we present the complete algorithm built with the above ingredients, and designed for converging to the continuous solution. In section 6.3, we compare the adapted mesh convergence with the classical uniform mesh convergence for an academic test case. Section 6.4 presents a case inspired by an industrial problem. Two other examples are presented in section 6.5.

6.2. A mesh convergence algorithm

The purpose of the mesh convergence algorithm is to build an idea of the continuous solution to the flow problem by proposing:

– several approximate solutions;

– a rather reliable estimate on the accuracy of the solutions.

This will be obtained by an algorithm producing a set of solutions showing convergence to the continuous limit with a constant satisfactory order. This algorithm will involve a flow kernel, an adaptative mesh generator and tools for comparing the solutions and certifying the convergence.

Since we want to get second-order convergence to the continuous solution with sequences of adapted very irregular meshes, it is compulsory to have a numerical approximation that will be actually second-order accurate in these conditions. We shall work with the scheme introduced in Chapter 1.

6.2.1. *Mesh adaptation with a fixed complexity*

For a given number of nodes, we want to specify an adapted mesh. We seek a mesh adaption method that satisfies the following properties:

i) For a given mesh size, represented by complexity, the method specifies essentially only one mesh and one discrete solution. By essentially, we mean that two meshes that produce about the same quality of approximation in all regions of the computational domain are just two representants of the same class of equivalence of meshes.

ii) For a sufficiently regular continuous solution, a model mesh description should be made up of a metric, which is independent of mesh complexity. Multiplying this metric by a scalar factor featuring mesh fineness will allow the generation of a series of *metric-embedded meshes*.

iii) For singular solutions, the mesh should enjoy second-order discontinuity capturing.

From the approximate flow solution, we extract a particular scalar quantity defined at every vertex on the mesh, the *sensor*. To fix the idea, we take the Mach number. We do not think that the Mach number makes the best sensor. Its replacement by a better option according to each test case deserves a special careful study. From an approximate Hessian matrix of the sensor, we derive the *metric*. Two options are studied: (i) in the isotropic case, the matrix is the product of the identity matrix by a scalar taken equal to the absolute value of the largest eigenvalue of the above Hessian. (ii) In the anisotropic case, the matrix is essentially the absolute value of the Hessian. Once the metrics are obtained, they are used as input in a *mesh regenerator*, for rebuilding a new mesh following the metrics, and involving approximatively the prescribed number of vertices.

For some 2D computations of this chapter, we applied the software BAMG, proposed by Hecht and co-workers (Castro-Díaz et al. 1999). BAMG uses the metrics defined on the initial mesh in order to test if some local improvement is necessary. Different local tools such as edge suppression and swapping, vertex suppression, addition, and reallocation are applied until the metrics are satisfied, that is, typically, every edge has a length close to unity for the considered metrics. A lot of 3D adaptive mesh generators are also now able to interpret the metric concept. Let us mention Frey (2001) for discrete surface mesh adaptation and (Bottasso 2004), Coupez (2000), Dobrzynski and Frey (2008), George (1999), Jones et al. (2006), Li et al. (2005), Pain et al. (2001), and Tam et al. (2000) in 3D. In the present chapter, we also use the generators of Dobrzynski and Frey (2008) and George (1999).

A crucial issue is the iteration and *nonlinear convergence of adaption* fixed point. In order to obtain the more or less unique mesh that is perfectly adapted according to our algorithm, the new adapted mesh is re-injected in an adaptation loop in order to get a new solution, then a new mesh, again with the same prescribed number of nodes. This algorithm is sketched in algorithm 5.1. of Chapter 5. Typically, after 6–10 cycles, the mesh and the solution do not change significantly and *the adaptation loop is declared as iteratively converged.* An accurate interpolation is applied to transfer solutions between the successive meshes in order to save computational effort. This tool is described in the sequel. The result of this process is to construct an approximate adapted solution for a prescribed number of nodes.

The *L^2-feature-based multiscale criterion* is chosen, with *the Mach number as the sensor field* and with double-projection as the recovery method (described in Chapter 5).

6.2.2. *Transfers and numerical convergence*

6.2.2.1. *Mesh convergence sequence*

We first propose to compute the numerical order from three approximate solutions for three prescribed different numbers of vertices: $N, 4N, 16N$. We note that in the case of adapted meshes, the meshes are not embedded. Further, the ratio between the vertices of two meshes does not need to be chosen as a multiple of 4. The choice of this ratio has just the advantage of an easy comparison with a classical embedded mesh convergence (uniform mesh division).

6.2.2.2. *Transfers to a unique mesh*

In order to compare solutions and to be able to compute their differences, we project all solutions on the finest mesh of the considered series. Two types of projection are tried and compared. The P_1 projector will keep exactly the P_1 interpolation of the coarse mesh after transfer to any finer mesh. Instead of this canonical option, and in order to have a transfer more robust in case of non-embedded meshes, we apply a projector which interpolates the function with a P_2 operator inside triangles (P_3 along edges) built from the knowledge of average gradients at vertices.

6.2.2.3. *Evaluation of the numerical order of convergence*

Numerical convergence evaluation has been formalized in (Roache 1994). The analysis proposed here is an adapted reformulation. We denote by u_h and u_{exact} the approximate and exact solutions, respectively. The αth-order convergence assumption is set for the L^2 space:

$$u_h = u_{exact} + h^\alpha u_1 + h^{\alpha+1} O_{L^2}(1), \qquad [6.1]$$

where $O_{L^2}(1)$ is a function which is bounded in L^2 norm when N tends to infinity. Let d be the dimension of computational space. For any mesh of N vertices, we denote by u_N the corresponding approximate solution. The model of analysis in terms of N is written as:

$$u_N(\mathbf{x},t) = u_{exact}(\mathbf{x},t) + N^{-\frac{\alpha}{d}}u_1(\mathbf{x},t) + o(N^{-\frac{\alpha}{d}}) \qquad [6.2]$$

In order to determine α, an estimation is done on three meshes of different sizes: N_1, N_2 and N_3 are the number of vertices and u_{N_1}, u_{N_2} and u_{N_3} the corresponding numerical solutions. Then we look for α such that:

$$\frac{1-\frac{N_2}{N_1}^{-\frac{\alpha}{d}}}{1-\frac{N_3}{N_1}^{-\frac{\alpha}{n}}} \approx \frac{u_{N_1}-u_{N_2}}{u_{N_1}-u_{N_3}}. \qquad [6.3]$$

In our examples, N_1 will represent the higher number of vertices and N_3 represents the lowest number of vertices. In practical computations, once the numerical order α is obtained, we shall say that the second-order convergence is approximatively observed when the numerical order lies between 1.6 and 2.2:

$$\alpha \in [1.6\ ,\ 2.2] \qquad [6.4]$$

6.2.2.4. *Global algorithm*

The standard way in mesh convergence is to pass from a coarser mesh of N nodes to a finer with $\eta^d N$ nodes where η is strictly larger than 1. It is given by Algorithm 6.1.

Algorithm 6.1. Convergent steady fixed-point: research of asymptotic convergence

1) Start with two adapted solutions $\mathcal{S}_N$ and $\mathcal{S}_{\eta^d N}$ with N and $\eta^d N$ as complexities or node numbers.

2) Compute an adapted solution $\mathcal{S}_{\eta^{2d} N}$ with $\eta^{2d} N$ as complexity or node number.

For this, iterate algorithm 5.1. of Chapter 5 until convergence.

3) Transfer all solutions to the finest mesh. Test whether the triplet of $(\mathcal{S}_N, \mathcal{S}_{\eta^d N}, \mathcal{S}_{\eta^{2d} N})$ satisfies the numerical order statement [6.4].

4) If not, set $N = \eta^d N$ and go to 1.

5) If yes, stop.

In the worst case, the different meshes are embedded. A simplified way to do this is to locally divide the edges in two equal edges[1], which induces that $\eta = 2$. In practice, this results in a number of cells (or d.o.f.) multiplied by four in 2D, eight in 3D, and necessitate to have three embedded meshes with $N, 2^d N, 2^{2d} N$ cells with second-order convergence on the three of them, this condition being difficult to obtain on the coarser one. In contrast, in the case of metric-governed adapted meshes, any η in $]0, 2[$ can be chosen. In 3D, choosing $\eta = 2^{1/3}$ is a reasonable option.

6.3. An academic test case

Obtaining a second-order convergence on three meshes with the above ratio, that is, $(N, 4N, 16N)$ nodes, is not a trivial task, even for very simple and smooth flows such as 2D airfoil flows at a very low Reynolds, $Re = 73$. The airfoil geometry is the standard NACA0012, and two farfield Mach numbers are considered, $M = 0.8$ and $M = 1.2$.

6.3.1. *Uniform refinement study*

A traditional mesh convergence study relies on a series of *embedded* meshes. It is the occasion to check how mesh convergence works with *unstructured* meshes. In order to have four meshes that are at the same time embedded and unstructured, we start from a fully unstructured coarse mesh and then the first intermediate mesh is obtained by *dividing each triangle into four equal subtriangles*. The third mesh is derived from the first intermediate one by dividing again each triangle in four equal triangles. The last (finest) mesh is in turn obtained by division of the previous one. The numbers of vertices (=nodes) of these meshes are 800, 3,114, 12,284, 48,792. By the way, we note that the new meshes are still unstructured, but they have lost the smoothness of the local mesh size, since this mesh size stays constant over macro-triangles and varies abruptly at the limits of the macro-triangles (the second-order derivative of the local mesh size is zero over macro-triangles, but is a Dirac at their limits). In the case of an elliptic finite-element study, this loss of regularity would not be an obstacle to second-order convergence and we assume this property extends to our case. For each computation, the equation residual is converged down to the zero machine. Figure 6.1 gives an idea of the quality of the various computations. We observe on the coarsest results some oscillations that indicate that the mesh is too coarse and irregular. We consider two sequences of three meshes:

a) meshes with (800, 3,114, 12,284) nodes, denoted by *mesh sequence (a)*;

b) meshes with (3,114, 12,284, 48,792) nodes, denoted by *mesh sequence (b)*.

1 In fact, the reasonable strategy is to start from an extremely fine mesh and to eliminate every other plan, line, node to get smoothly meshed embedded grids.

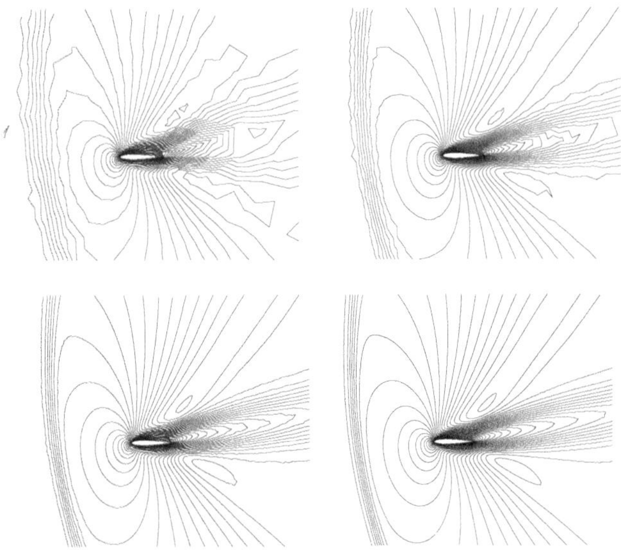

Figure 6.1. *Uniform refinement: pressure contours for the supersonic flow case with embedded meshes of 800, 3,114, 12,284 and 48,792 vertices (from left to right and top to bottom)*

We present in Table 6.1 the numerical convergence orders evaluated as explained in section 6.2.

Variables	Mach 0.8 sequence (a)	Mach 0.8 sequence (b)	Mach 1.2 sequence (a)	Mach 1.2 sequence (b)
Rho	1.124	1.984	0.948	1.139
Rho*U	0.889	1.793	1.016	1.851
Rho*V	0.780	1.716	1.024	1.121

Table 6.1. *NACA0012 test case. Uniform refinement: several measures of L^2 convergence order, measured on the finer mesh. Sequence (a) involves meshes with (800, 3,114, 12,284) nodes, sequence (b) involves meshes with (3,114, 12,284, 48,792) nodes*

All the convergence studies on the sequence of coarsest meshes (a) produce a convergence order around 1. This means that not all of the three calculations are in the asymptotic region where dividing the mesh step by two ensures a four times smaller error. Convergence orders for the $M = 0.8$ case with the finest meshes (b) appear to be near 2. For the supersonic case, convergence order is near 1, even with the sequence of finest meshes (except for the horizontal velocity where we get the second order). We attribute this difference to the influence of the bow shock, which is not well captured on the different meshes. To sum up, the second-order accuracy of the approximation scheme on unstructured meshes is confirmed, but the above sequence of meshes shows that only fine meshes give some access to second-order accuracy and this only for solutions without viscous shock.

6.3.2. *Isotropic adaptation study*

Now the computations are done by applying the mesh adaptation process described in algorithm 6.1. A unique sensor, the Mach number is used, and the maximum eigenvalue of its local Hessian is used in the isotropic metric. The iteration between computation and mesh regeneration applied until convergence is fully obtained between mesh and solution (typically 10 times). The refinement factor η is taken equal to 2 in terms of local mesh size, which means that as for the previous uniform refinement case, the number of vertices is about four times larger at each refinement. We observe that with the sequence of finest meshes, a second-order convergence is reached for the two farfield Mach numbers. We recall that from one case to the other one, two adapted meshes with an equivalent fineness level are different since they are adapted to different flows. However, second-order convergence is not observed with the coarse sequences, except for the horizontal momentum (Table 6.2).

Variables	Mach 0.8 sequence (a)	Mach 0.8 sequence (b)	Mach 1.2 sequence (a)	Mach 1.2 sequence (b)
Rho	1.05	1.78	1.00	2.13
Rho*U	1.87	1.89	1.61	1.55
Rho*V	1.26	1.71	0.99	1.80

Table 6.2. *NACA0012 test case. Isotropic adaptative refinement: several measures of L^2 convergence order*

6.3.3. *Anisotropic adaptation study*

Algorithm 6.1. is again applied with $\eta = 2$. The absolute value of the local Mach number Hessian is used in the anisotropic metric. From Table 6.3, we deduce that, with the anisotropic refinement, second-order convergence is obtained even with the coarse sequences of refinement for both cases, subsonic and supersonic, and for the three considered variables. This convergence is obtained with meshes involving a number of vertices that is four times smaller than with the previous strategies. This ratio depends on the Reynolds number and would be much larger for thinner shocks.

REMARK 6.1.– As mentioned earlier, we have not considered the total energy convergence, and in fact, with the Mach number as adaptative sensor, we do not get for the energy a convergence as good as for the three other flow variables. However, a good convergence of lift and drag coefficients can be obtained with this sensor, and examples are given in Chapter 5 of Volume 2.

Variables	Mach 0.8 sequence (a)	Mach 0.8 sequence (b)	Mach 1.2 sequence (a)	Mach 1.2 sequence (b)
Rho	1.74	2.01	1.62	2.15
Rho*U	1.58	2.08	1.78	1.75
Rho*V	1.83	1.66	1.85	1.87

Table 6.3. *NACA0012 test case. Anisotropic adaptative refinement: several measures of L^2 convergence order*

6.3.4. *Error level*

The validation of second-order accuracy, that is the maximal theoretical order of accuracy, allows to derive an error estimate of the L^2 error for the finest mesh among the coarse sequences. Let us denote by u_2 the solution computed with the adapted anisotropic mesh of 3,114 vertices and u_3 the solution computed with the adapted anisotropic mesh of 11,938 vertices for the supersonic case. Since we have second-order convergence, and starting from the deviation:

$$||u_2 - u_3||_{L^2} \;=\; 1,802\;10^{-4}\;,$$

we derive the following *error estimate*:

$$||u_3 - u||_{L^2} \;\leq\; (1,802/3)\;10^{-4} = 0,600\;10^{-4}\;. \qquad [6.5]$$

As can be expected, this estimate is coherent with the calculation of u_4 done with the finest adapted anisotropic mesh. Indeed, we have:

$$||u_3 - u_4||_{L^2} \;=\; 5,637\;10^{-5}$$

which confirms the accuracy of [6.5]. We observe that the error estimate for u_4 would be of 2×10^{-5}, a rather small error for a second-order accurate computation.

As a conclusion, this analysis gives some insurance that second-order convergence is reached, “certified”, and that the approximation error is reasonably well predicted by the convergence model.

6.4. 3D multiscale anisotropic mesh adaptation

In this section, we study a supersonic jet geometry provided by Dassault Aviation. The geometry is shown in Figure 6.3[2]. The multiscale mesh adaptation considers a control of the interpolation error in $\mathbf{L}^2$ norm of the local Mach number. The local Mach number has been selected as it is really representative of supersonic flows. We have deliberately avoided to adapt the aircraft surface mesh. Indeed, the provided surface mesh is very accurate and is of high quality for the computation. A mesh gradation of 2.5 has been set (Alauzet 2009). Algorithm 6.1. is applied with $\eta = 2^{1/3}$. A total of 32 adaptations have been performed. The mesh adaptation loop is split into four steps of eight adaptations. At each step, the couple mesh-solution is algorithmically converged at a fixed metric complexity. With the choice $\eta = 2^{1/3}$, this complexity is multiplied by two between two steps. This strategy has two main advantages. First, it enables us to perform a convergence study of the whole solution as a series of couple mesh-solution with an increasing accuracy is obtained. Second, considering an increasing dynamic complexity level accelerates the convergence of the whole process. We work then in terms of complexities, which we fix to

$$[100{,}000,\ 200{,}000,\ 400{,}000,\ 800{,}000],$$

which give meshes the sizes of which are about 0.8, 1.7, 4 and 9 millions of vertices. To validate the CFD computations, the solution global convergence order is analyzed with respect to a reference solution. The reference solution is the final solution obtained on the finest adapted mesh for a complexity equal to $800,000$. The convergence is computed in $\mathbf{L}^2$ norm on the local Mach number which is the variable used for the mesh adaptation. A convergence order equal to 2 is obtained as predicted by the theory (see Figure 6.2, left). With regard to the obtained solution, a very accurate result in the whole computational domain is achieved for the final adapted meshes containing almost 9 millions vertices. In particular, the accuracy of the tetrahedral meshes has reached the surface meshes accuracy and the refinements have been propagated in the whole computational domain. This is illustrated in Figure 6.2 (right), where the Mach cones of the jet is depicted. This result points out that the numerical dissipation of the flow solver has been drastically reduced because of the anisotropic mesh adaptation. The shock waves have been accurately propagated in the whole computational domain, more than 1 km below the jet. Figures 6.3 and 6.4

2 The length of the jet is $L = 37$ m, and it has a wing span of 17 m. The surface mesh accuracy varies between 1 mm and 30 cm. The computational domain is a cylinder of 2.25 km length and 1.5 km diameter. This represents a scale factor of 10^6 if the size of the domain is compared to the maximal accuracy of the jet surface mesh. The flight conditions are a supersonic cruise speed of Mach 1.6 at an altitude of 13,680 m (45,000 feet). In this study, the cruise lift has been set to $C_l = 0.115$. The angle of attack for the aircraft is set such that this lift is attained. The obtained angle of attack is near to 3°.

show the final adapted mesh (middle, in the symmetry plane Oxz and, bottom, in the cut plane Oxy) and the associated local Mach number iso-values. We note that the shock waves and the anisotropic refinements have been propagated in the whole domain without a notable dissipation. We also remark the very accurate capture inside the mesh of all the shock waves. This points out the multiscale behavior of the mesh adaptation approach. Each element of the geometry emits its own shock waves. The focalization of shocks during their propagation is also illustrated inside the mesh. These pictures also point out the great complexity of the physical phenomenon and the mesh refinement. This demonstrates the necessity to use a fully automatic mesh adaptive method.

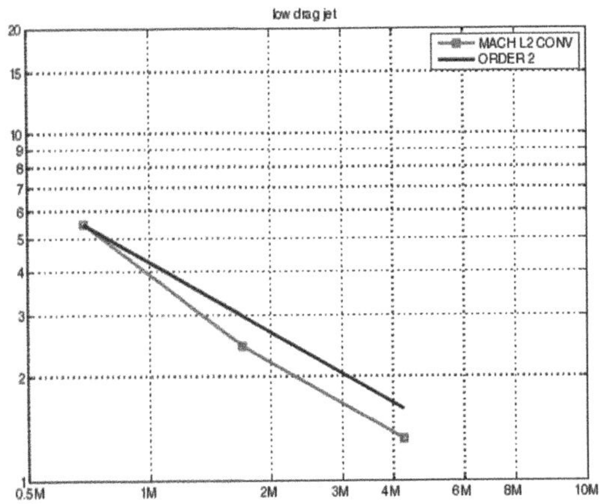

Figure 6.2. *Left: Global mesh convergence order for the supersonic jet: a second-order spatial convergence for* $||M - M_{ref}||$ *is obtained where* M_{ref} *holds for a reference calculation. Right: The Mach cone emitted by the jet. The maximal Mach cone diameter is 1.25 km. For a color version of this figure, see www.iste.co.uk/dervieux/meshadaptation1*

6.5. Conclusion

The method for obtaining second-order numerical convergence with a metric-based mesh adaption described in this chapter uses two important ideas: (1) the mesh adaption has to be governed by a metrics (continuous mesh approach), and (2) fixed points coupling mesh and solution are converged for each level of complexity.

Figure 6.3. *Left: Final surface mesh for the supersonic jet computation containing 9,465,835 vertices and 56,568,966 tetrahedra. Right: The associated local Mach number iso-value. For a color version of this figure, see www.iste.co.uk/dervieux/meshadaptation1*

This allows early discontinuity and steep gradient second-order capturing. The method relies on a set of tools dedicated to produce "a certified numerical simulation". Starting from a first tool, an approximation enjoying a good behavior on irregular meshes, we have combined it with a second tool, an adaptation loop. An important point is that mesh adaptation is primarily performed for a fixed number of nodes. The complexity is increased in a outer loop. Mesh convergence is measured. The third set of tools deals with the transfers of solutions between meshes and the solutions comparison for the certification of the convergence rate to the continuous limit.

With a first 2D example, we have seen that second-order convergence with uniform refinement is not easily obtained even with a rather large number of nodes, while, with the proposed anisotropic mesh adaptation approach, second-order convergence is observed with a much smaller number of nodes. We have shown that the proposed method applies successfully to a more complex CFD calculation. In the case of high Reynolds turbulent flows, the uniform refinement does not produce any reasonable result. The isotropic option is clearly inaccurate.

Although less rigorous than an exact *a posteriori* error estimate, the proposed mesh convergence analysis is a rather reliable method and carries some security in the use of the computed results. The cost of our computational method is not very high, since the coarse adapted meshes produce already rather accurate solutions, and we can use them as initial conditions on finer mesh.

Choosing, as we did, a clever sensor (like Mach number) is an option, which lacks a firm theoretical basis. In computations, this option appeared to be not adequate for total energy and pressure convergence. A more conservative strategy is to introduce all EDP variables in the sensor (with metric intersection). Error models relying on the new methods of adjoint states for mesh adaptation are addressed in another chapter.

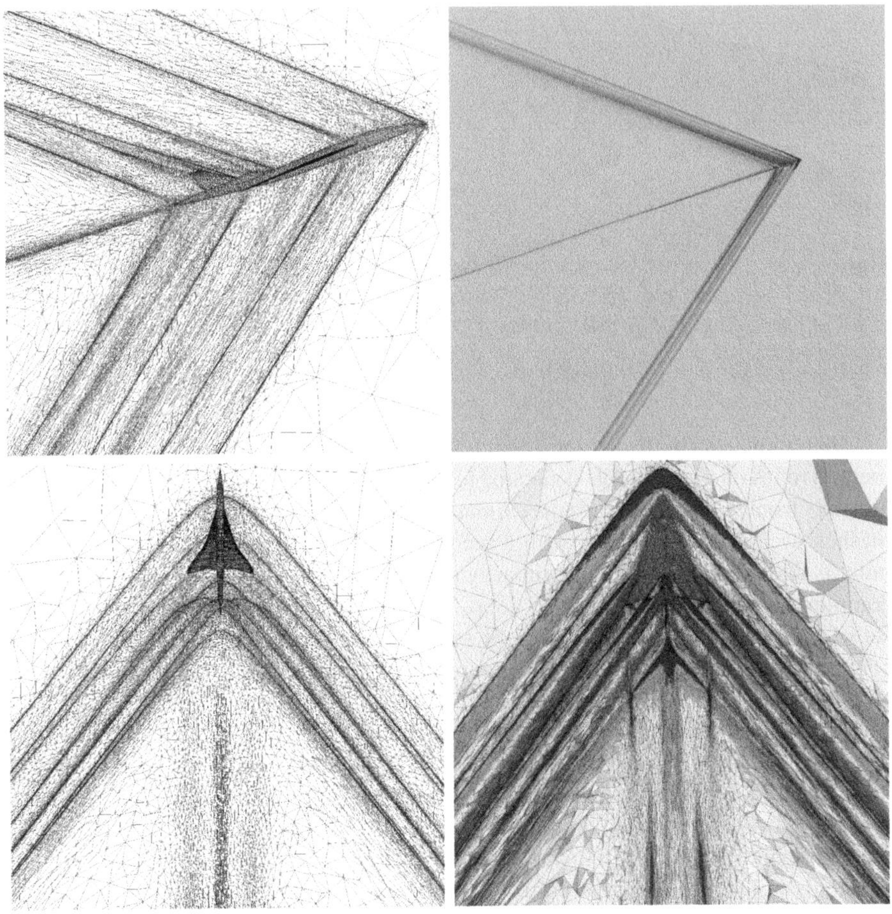

Figure 6.4. *Left: Final adapted mesh for the supersonic jet computation containing 9,465,835 vertices and 56,568,966 tetrahedra. Right: The associated local Mach number iso-value. From top to bottom, the cut plane along Oxz and view of the adapted volume mesh through the cut plane along Oxy. For a color version of this figure, see www.iste.co.uk/dervieux/meshadaptation1*

6.6. Notes

Notable contributions to mesh adaptation have taken into account the whole process from a numerical scheme, local errors and particularly mesh specifications.

We refer, for example, to Becker et al. (1999); Giles and Suli (2002). Since an adapted mesh can be specified, this may mean that the mesh satisfying some specifications is somewhat unique in a class of equivalent meshes. Several teams have adopted this line of reasoning (see, for example, Castro-Díaz et al. 1999 and Habashi et al. 2000). After being specified, the adapted mesh must be built. This can be done by taking benefit from the recent progress in the elaboration of anisotropic meshes, either by a powerful improvement of the current mesh (Habashi et al. 2000) or by its regeneration (Castro-Díaz et al. 1999) according to the given metrics.

Another efficiency issue. A question not addressed in this chapter concerns possible ways to control accurately the effort in iteratively solving the discrete equation between two mesh adaptations. Given a mesh, we call the approximation error the deviation to exact solution after iterative convergence. It is generally admitted that when iteratively converging:

1) during a first part of the convergence, the iterative error is larger than the approximation error;

2) the two errors are of same order of magnitude;

3) in a third part of the convergence, iterative error is much smaller, indicating that the iterations of the third part are useless.

To control this process, it is necessary to evaluate once an approximation error and to evaluate at each solver main iteration an iterative error comparable to the approximation error in order to be able to decide to stop convergence by comparing them. An example is given in Brèthes et al. (2015). When the adaptation algorithm involves an evaluation of a corrector, as defined in Chapter 1 of Volume 2, this corrector can be used.

Extrapolation. Once convergence at some order is verified, a Richardson extrapolation can be tried. In principle, the extrapolation of three second-order accurate calculations will produce a third accurate extrapolate. However, the convergence is generally true for integral outputs depending om the solution field, but not for a pointwise value in the field itself.

Final error estimate. The same type of remark applies to the estimation of the error. Richarson analysis can produce reliable error estimate for integral outputs, not for pointwise values.

Although this is not theoretically established, the corrector tool described in Chapter 1 of Volume 2 is in our opinion likely to produce a better estimate for pointwise values.

References

Abalakin, I., Bobkov, V., Dervieux, A., Kozubskaya, T., Shiryaev, V. (2001). Study of high accuracy order schemes and non-reflecting boundary conditions for noise propagation problems. Liapunov Institute Report, project 99-02 [Online]. Available at: http://www-sop.inria.fr/tropics/Alain.Dervieux/liapunov.html.

Abalakin, I., Dervieux, A., Kozubskaya, T. (2002a). A vertex centered high order MUSCL scheme applying to linearised Euler acoustics. Research report RR-4459, INRIA.

Abalakin, I., Dervieux, A., Kozubskaya, T. (2002b). Computational study of mathematical models for noise DNS. AIAA Paper 2002–2585.

Abalakin, I., Dervieux, A., Kozubskaya, T. (2002c). High accuracy study of mathematical models for DNS of noise around steady mean flow. In *Proc. West East High Speed Flow Field Conference 2002*, Zeitoun, D.E., Périaux, J., Desideri, J.A., Marini, M. (eds). CIMNE, Barcelona.

Abalakin, I., Dervieux, A., Kozubskaya, T. (2004). On accuracy of noise direct calculation based on Euler model. *Int. J. Acoustics*, 3(2), 157–180.

Abgrall, R. (1994). On essentially non-oscillatory schemes on unstructured meshes, analysis and implementation. *J. Comp. Phys.*, 114, 45–58.

Abgrall, R. (1996). How to prevent pressure oscillations in multicomponent flows: A quasi conservative approach. *J. Comp. Phys.*, 125, 150–160.

Abgrall, R., Dobrzynski, C., Froehly, A. (2012). A method for computing curved 2D and 3D meshes via the linear elasticity analogy: Preliminary results. Research report 8061, INRIA.

Alauzet, F. (2009). Size gradation control of anisotropic meshes. *Finite Elem. Anal. Des.*, 46, 181–202.

Alauzet, F. and Frey, P.J. (2003). Estimateur d'erreur géométrique et métrique anisotropes pour l'adaptation de maillage. Partie I : aspects théoriques. RR-4759, INRIA [in French].

Alauzet, F. and Loseille, A. (2009). On the use of space filling curves for parallel anisotropic mesh adaptation, In *Proc. 18th International Meshing Roundtable*, Springer, Berlin.

Alauzet, F. and Loseille, A. (2010). High order sonic boom modeling by adaptive methods. *J. Comput. Phys.*, 229, 561–593.

Alauzet, F. and Mehrenberger, M. (2010b). P1-conservative solution interpolation on unstructured triangular meshes. *Int. J. Numer. Meth. Eng.*, 84(13), 1552–1588.

Alauzet, F., Loseille, A., Dervieux, A., Frey, P.J. (2006). Multi-dimensional continuous metric for mesh adaptation. In *Proc. 15th International Meshing Roundtable*, Springer, Berlin.

Alauzet, F., Frey, P.J., George, P.-L., Mohammadi, B. (2007). 3D transient fixed point mesh adaptation for time-dependent problems: Application to CFD simulations. *J. Comp. Phys.*, 222, 592–623.

Alcin, H., Koobus, B., Allain, O., Dervieux, A. (2013). Efficiency and scalability of a two-level Schwarz algorithm for incompressible and compressible flows. *Int. J. Numer. Methods Fluids*, 72(1), 69–89.

Allain, O., Guégan, D., Alauzet, F. (2009). Studying the impact of unstructured mesh adaptation on free surface flow simulations. In *Proceedings of the ASME 28th International Conference on Ocean, Offshore and Arctic Engineering*, Honolulu, Hawaii.

Anderson, W.K. and Bonhaus, D.L. (1994). An implicit upwind algorithm for computing a turbulent flow on unstructured grids. *Comput. Fluids*, 23(1), 1–21.

Anglada, M.V., Garcia, N.P., Crosa, P.B. (1999). Directional adaptive surface triangulation. *Comput. Aided Des.*, 16, 107–126.

Arsigny, V., Fillard, P., Pennec, X., Ayache, N. (2006). Log-Euclidean metrics for fast and simple calculus on diffusion tensors. *Magn. Reson. Med.*, 56(2), 411–421.

Baba, K. and Tabata, M. (1981). On a conservative upwind finite element scheme for convective diffusion equations. *R.A.I.R.O Numer. Anal.*, 15, 3–25.

Baines, M.J. (1994). *Moving Finite Elements*. Oxford University Press, Inc., New York.

Baker, T.J. (1997). Mesh adaptation strategies for problems in fluid dynamics. *Finite Elem. Anal. Des.*, 25, 243–273.

Bakhvalov, P., Abalakin, I., Kozubskaya, T. (2014). Edge-based reconstruction schemes for prediction of near field flow region in complex aeroacoustic problems. *Int. J. Aeroacoustics*, 13(3–4), 207–234.

Bank, R.E. and Smith, R.K. (1997). Mesh smoothing using a posteriori error estimates. *SIAM J. Numer. Anal.*, 34(3), 979–997.

Barth, T.J. (1994). Aspects of unstructured grids and finite volume solvers for the Euler and Navier-Stokes equations. *Computational Fluid Dynamics*, 1994–04 in Lecture Series, von Karman Institute for Fluid Dynamics.

Batina, J.T. (1989). Unsteady Euler algorithm with unstructured dynamic mesh for complex-aircraft aeroelastic analysis. In *AIAA 30th Structures, Structural Dynamics and Materials Conference*, AIAA-1989-1189, Mobile, AL.

Batten, P., Clarke, N., Lambert, C., Causon, D.M. (1997). On the choice of wavespeeds for the HLLC Riemann solver. *SIAM J. Sci. Comput.*, 18(6), 1553–1570.

Becker, R., Braack, M., Rannacher, R. (1999). Numerical simulation of laminar flames at low Mach number with adaptative finite elements. *Combus. Theory Model.*, 3, 503–534.

Belhamadia, Y., Fortin, A., Chamberland, E. (2004). Three-dimensional anisotropic mesh adaptation for phase change problems. *J. Comp. Phys.*, 201, 753–770.

Berger, M. (1987a). *Geometry 1*. Springer Verlag, Berlin.

Berger, M. (1987b). *Geometry 2*. Springer Verlag, Berlin.

Berger, M. (2003). *A Panoramic View of Riemannian Geometry*. Springer Verlag, Berlin.

Berger, M. and Colella P, P. (1989). Local adaptive mesh refinement for shock hydrodynamics. *J. Comp. Phys.*, 82(1), 67–84.

Berzins, M. (1998). A solution based triangular and tetrahedral mesh quality indicator. *SIAM J. Sci. Comput.*, 19(6), 2051–2060.

Binev, P., Dahmen, W., DeVore, R. (2004). Adaptive finite element methods with convergence rates. *Numer. Math.*, 97, 219–268.

Borouchaki, H. and George, P.-L. (1996). Maillage de surfaces paramétriques. Partie i : aspects théoriques. Technical report no. 2928, INRIA Rocquencourt.

Borouchaki, H. and George, P.-L. (2017). *Meshing, Geometric Modeling and Numerical Simulation 1: Form Functions, Triangulations and Geometric Modeling*. ISTE Ltd, London, and John Wiley & Sons, New York.

Botella, O. and Peyret, R. (1998). Benchmark spectral results on the lid-driven cavity flow. *Comput. Fluids*, 27, 421–433.

Bottasso, C.L. (2004). Anisotropic mesh adaption by metric-driven optimization. *Int. J. Numer. Meth. Eng.*, 60, 597–639.

Brèthes, G., Allain, O., Dervieux, A. (2015). A mesh-adaptative metric-based full-multigrid for the Poisson problem. *Int. J. Numer. Methods Fluids*, 79(1) [Online]. Available at: http://www-sop.inria.fr/members/Gautier.Brethes/article-ADA-MG.pdf.

Bruchon, J., Digonnet, H., Coupez, T. (2009). Using a signed distance function for the simulation of metal forming process: Formulation of the contact condition and mesh adaptation. From Lagrangian approach to an Eulerian approach. *Int. J. Numer. Methods Eng.*, 78(8), 980–1008.

Buscaglia, G.C. and Dari, E.A. (1997). Anisotropic mesh optimization and its application in adaptivity. *Int. J. Numer. Methods Eng.*, 40, 4119–4136.

Cabello, J., Lohner, R., Jacquotte, O.-P. (1991). A variational method for the optimization of directionally stretched elements generated by the advancing front method (AFM). In *Proceedings of the 3rd International Grid Conference on Numerical Grid Generation in Computational Field Simulation and Related Fields*, Arcilla, A.S., Auser, J.H., Eiseman, P.R., Thompson, J.F. (eds). North Holland, Amsterdam.

Cai, X.-C. and Sarkis, M. (1999). A restricted additive Schwarz preconditioner for general sparse linear systems. *SIAM J. Sci. Comput.*, 21(2), 792–797.

Camarri, S., Salvetti, M.-V., Koobus, B., Dervieux, A. (2001). Numerical diffusion based on high-order derivatives in MUSCL schemes for LES on unstructured grids. In *Proceedings of DLES-4, Direct and Large-Eddy Simulation-IV*, Twente, The Netherlands.

Camarri, S., Salvetti, M.-V., Dervieux, A., Koobus, B. (2002). A low diffusion MUSCL scheme for LES on unstructured grids. Technical report RR-4412, INRIA.

Camarri, S., Koobus, B., Salvetti, M.-V., Dervieux, A. (2004). A low diffusion MUSCL scheme for LES on unstructured grids. *Comput. Fluids*, 33, 1101–1129.

Carabias, A., Belme, A., Loseille, A., Dervieux, A. (2018). Anisotropic goal-oriented error analysis for a third-order accurate CENO euler discretization. *Int. J. Numer. Methods Fluids*, 86(6), 392–413.

do Carmo, M. (1976). *Differential Geometry of Curves and Surfaces*. Prentice Hall, Upper Saddle River, NJ.

Carpenter, M.H. and Casper, J.H. (1999). Accuracy of shock capturing in two spatial dimensions. *AIAA J.*, 37(9), 1072–1079.

Castro-Dìaz, M.J., Hecht, F., Mohammadi, B., George, P.-L. (1996). Anisotropic adaptative mesh generation in two dimensions for CFD. In *Computational Fluids Dynamic '96*, Paris, France, Wiley, Chichester.

Castro-Díaz, M.J., Hecht, F., Mohammadi, B., Pironneau, O. (1997). Anisotropic unstructured mesh adaptation for flow simulations. *Int. J. Numer. Methods Fluids*, 25, 4, 475–491.

Castro-Díaz, M.J., Borouchaki, H., George, P.-L., Hecht, F., Mohamadi, B. (1999). Anisotropic adaptative mesh generation in two dimensions for CFD. In *Numerical Methods in Engineering '96: Proceedings of the Second ECCOMAS Conference on Numerical Methods in Engineering*, 9–13 September 1996.

Catalano, L.A. (2002). A new reconstruction scheme for the computation of inviscid compressible flows on 3D unstructured grids. *Int. J. Numer. Methods Fluids*, 40(1–2), 273–279.

Catris, S. and Aupoix, B. (2000). Density corrections for turbulence models. *Aerosp. Sci. Technol.*, 4, 1–11.

Chen, L., Sun, P., Xu, J. (2007). Optimal anisotropic meshes for minimizing interpolation errors in L^p-norm. *Math. Comput.*, 76(257), 179–204.

Ciarlet, P.G. (1978). *The Finite Element Method for Elliptic Problems*. North-Holland Publishing Company, Amsterdam.

Ciarlet, P.G. (1991). Basic error estimates for elliptic problems. In *Handbook of Numerical Analysis II*, Ciarlet, P.G. and Lions, J.L. (eds). North-Holland, Amsterdam.

Ciarlet, P.G. (2005). *An Introduction to Differential Geometry, with Applications to Elasticity*. Springer, Dordrecht.

Clément, P. (1975). Approximation by finite element functions using local regularization. *Revue française d'automatique, informatique et recherche opérationnelle*, R-2, 77–84.

Cockburn, B. (2003). Discontinuous Galerkin methods. *Z. Angew. Math. Mech.*, 83(11), 731–754.

Cockburn, B. and Shu, C.-W. (1989). TVB Runge-Kutta local projection discontinuous Galerkin finite element method for conservation laws II: General framework. *Math. Comput.*, 52, 411–435.

Compère, G., Marchandise, E., Remacle, J.-F. (2007). Transient adaptivity applied to two-phase incompressible flows. *J. Comp. Phys.*, 227, 1923–1942.

Compère, G., Remacle, J.-F., Jansson, J., Hoffman, J. (2010). A mesh adaptation framework for dealing with large deforming meshes. *Int. J. Numer. Methods Eng.*, 82, 843–867.

Coudière, Y., Palmerio, B., Dervieux, A., Leservoisier, D. (2002). Accuracy barriers in mesh adaptation. Research report RR-4528, INRIA.

Coupez, T. (2000). Génération de maillages et adaptation de maillage par optimisation locale. *Rev. Eur. Élém. Finis*, 9, 403–423.

Cournède, P.-H., Koobus, B., Dervieux, A. (2006). Positivity statements for a mixed-element-volume scheme on fixed and moving grids. *Eur. J. Comput. Mech.*, 15(7–8), 767–798.

Courty, F., Leservoisier, D., George, P.-L., Dervieux, A. (2006). Continuous metrics and mesh adaptation. *Appl. Numer. Math.*, 56(2), 117–145.

Debiez, C. (1996). Approximation et linéarisation d'écoulements aérodynamiques instationnaires. PhD Thesis, University of Nice [in French].

Debiez, C. and Dervieux, A. (1999). Mixed element volume MUSCL methods with weak viscosity for steady and unsteady flow calculation. *Comput. Fluids*, 29, 89–118.

Debiez, C., Dervieux, A., Mer, K., NKonga, B. (1998). Computation of unsteady flows with mixed finite/finite element upwind methods. *Int. J. Numeri. Methods Fluids*, 27, 193–206.

Deconinck, H., Roe, P.L., Struijs, R. (1993). A multidimensional generalization of Roe's flux difference splitter for the Euler equations. *Comput. Fluids*, 22(23), 215–222.

Dervieux, A. (1985). Steady Euler simulations using unstructured meshes. *von Karman Institute for Fluid Dynamics*. Lecture Series 1985-04.

Dervieux, A. (1987). Steady Euler simulations using unstructured meshes. In *Partial Differential Equations of Hyperbolique Type and Applications*, Geymonat, G. (ed.). World Scientific, Hackensack, NJ.

Dervieux, A. and Thomasset, F. (1981). Multifluid incompressible flows by a finite element method. *Lecture Notes in Physics*, 11, 158–163.

Dobrzynski, C. and Frey, P.J. (2008). Anisotropic Delaunay mesh adaptation for unsteady simulations. In *Proc. 17th International Meshing Roundtable*, Springer, Berlin.

Dompierre, J., Vallet, M.G., Fortin, M., Bourgault, Y., Habashi, W.G. (1997). Anisotropic mesh adaptation: Towards a solver and user independent CFD. AIAA Paper, 97-0861.

Donea, J., Quartapelle, L., Selmin, V. (1987). An analysis of time discretization in the finite element solution of hyperbolic problems. *J. Comp. Phys.*, 70, 463–499.

Eça, L., Hoekstra, M., Hay, A., Pelletier, D. (2007). Verification of RANS solvers with manufactured solutions. *Eng. Comput.*, 23(4), 253–270.

Farhat, C. (1995). High performance simulation of coupled nonlinear transient aeroelastic problems. Special course on parallel computing in CFD. NATO AGARD Report, R-807.

Farhat, C. and Lesoinne, M. (2000). Two efficient staggered procedures for the serial and parallel solution of three-dimensional nonlinear transient aeroelastic problems. *Comput. Methods Appl. Mech. Eng.*, 182, 499–516.

Farhat, C., Degand, C., Koobus, B., Lesoinne, M. (1998). Torsional springs for two-dimensional dynamic unstructured fluid meshes. *Comput. Methods Appl. Mech. Eng.*, 163(5), 231–245.

Farhat, C., Geuzaine, P., Grandmont, C. (2001). The discrete geometric conservation law and the nonlinear stability of ALE schemes for the solution of flow problems on moving grids. *J. Comp. Phys.*, 174, 669–694.

Fezoui, L. (1985). Résolution des équations d'Euler par un schéma de Van Leer en éléments finis. Research report RR-0358, INRIA.

Fezoui, L. and Dervieux, A. (1989). Finite-element non oscillatory schemes for compressible flows. In *Symposium on Computational Mathematics and Applications 730*, Publications of University of Pavie.

Fezoui, L. and Stoufflet, B. (1989). A class of implicit schemes for Euler simulations with unstructured meshes. *J. Comp. Phys.*, 84(1), 174–206.

Formaggia, L. and Perotto, S. (2001). New anisotropic a priori error estimates. *Numer. Math.*, 89, 641–667.

Formaggia, L., Micheletti, S., Perotto, S. (2004). Anisotropic mesh adaptation in computational fluid dynamics: Application to the advection-diffusion-reaction and the Stokes problems. *Appl. Numer. Math.*, 51(4), 511–533.

Fortin, M. (2000a). Étude numérique d'estimations d'erreur a posteriori. *Rev. Eur. Élém. Finis*, 9(4), 467–486.

Fortin, M. (ed.) (2000b). Estimations a posteriori et adaptation de maillages. *Rev. Eur. Élém. Finis*, 9(Special Issue), 4.

Fortin, M., Vallet, M.-G., Dompierre, J., Bourgault, Y., Habashi, W.G. (1999). Anisotropic mesh adaptation: Theory, validation and applications. In *Numerical Methods in Engineering '96: Proceedings of the Second ECCOMAS Conference on Numerical Methods in Engineering*, 9–13 September 1996.

Fournier, L., Carré, G., Lanteri, S. (1998). Parallel multigrid algorithms for the acceleration of compressible flow calculations. *Calculateurs parallèles, réseaux et systèmes répartis*, 10(4), 381–389.

Francescatto, J. and Dervieux, A. (1998). A semi-coarsening strategy for unstructured multigrid based on agglomeration. *Int. J. Numer. Methods Fluids*, 26, 927–957.

Frey, P.J. (2001). Yams, a fully automatic adaptive isotropic surface remeshing procedure. Research report RT-0252, INRIA.

Frey, P.J. and Alauzet, F. (2005). Anisotropic mesh adaptation for CFD computations. *Comput. Methods Appl. Mech. Eng.*, 194(48–49), 5068–5082.

Frey, P. and George, P.-L. (1999). *Maillages*. Application aux elements finis, Hermes, Paris.

Frey, P.J. and George, P.-L. (2008). *Mesh Generation: Application to Finite Elements*, 2nd edition. ISTE Ltd, London, and John Wiley & Sons, New York.

George, P.-L. (1999). Tet meshing: Construction, optimization and adaptation. In *Proceedings of the 8th International Meshing Roundtable*, South Lake Tao, CA.

George, P.-L., Hecht, F., Vallet, M.G. (1991). Creation of internal points in Voronoi's type method. Control adaptation. *Adv. Eng. Software*, 13(5–6), 303–312.

George, P.-L. (2001). `Gamhic3d`, a fully automatic adaptive mesh generation method in three dimensions. Technical note, INRIA.

George, P.-L. (2003). `Gamanic3d`, adaptive anisotropic tetrahedral mesh generator. Technical note, INRIA.

George, P.-L., Borouchaki, H., Alauzet, F., Laug, P., Loseille, A., Maréchal, L. (2019). *Meshing, Geometric Modeling and Numerical Simulation 2: Metrics, Meshes and Mesh Adaptation*. ISTE Ltd, London, and John Wiley & Sons, New York.

George, P.-L., Alauzet, F., Loseille, A., Maréchal, L. (2020). *Meshing, Geometric Modeling and Numerical Simulation 3: Storage, Visualization and In Memory Strategies*. ISTE Ltd, London, and John Wiley & Sons, New York.

Giles, M.B. and Suli, E. (2002). Adjoint methods for PDEs: A posteriori error analysis and postprocessing by duality. *Acta Numerica*, 11, 145–236.

Gnoffo, P.A. (1983). A finite volume adaptive grid algorithm applied to planetary entry flowfields. *AIAA J.*, 21, 1249–1254.

Godlewski, E. and Raviart, P.-A. (1996). *Numerical Approximation of Hyperbolic Systems of Conservation Law*. Springer, Cham.

Goldberg, M. (1974). Three infinite families of tetrahedral space-fillers. *J. Comb. Theory, Ser. A*, 16(3), 348–354.

Gourvitch, N., Roge, G., Abalakin, I., Dervieux, A., Kozubskaya, T. (2004), A tetrahedral-based superconvergent scheme for aeroacoustics. Research report 5212, INRIA [Online]. Available at: http://www.inria.fr/RRRT/RR-5212.html

Gruau, C. and Coupez, T. (2005). 3D tetrahedral, unstructured and anisotropic mesh generation with adaptation to natural and multidomain metric. *Comput. Methods Appl. Mech. Eng.*, 194(48–49), 4951–4976.

Habashi, W.G., Dompierre, J., Bourgault, Y., Ait-Ali-Yahia, D., Fortin, M., Vallet, M.-G. (2000). Anisotropic mesh adaptation: Towards user-independent, mesh-independent and solver-independent CFD solutions: Part I: General principles. *Int. J. Numer. Methods Fluids*, 32, 725–744.

Harlow, F.H. and Welch, J.E. (1965). Numerical calculation of time-dependent viscous incompressible flow of fluid with a free surface. *Phys. Fluids*, 8, 2182–2189.

Harten, A. (1983). High resolution method for hyperbolic conservation laws. *J. Comput. Phys.*, 49, 357–393.

Harten, A. and Osher, S. (1987). Uniformly high order accurate nonoscillatory schemes. *SIAM J. Numer. Anal.*, 24, 279–309.

Harten, A., Lax, P.D., van Leer, B. (1983). On upstream differencing and Godunov-type schemes for hyperbolic conservation laws. *SIAM Revue*, 25(1), 35–61.

Hecht, F. (1998). BAMG: Bidimensional anisotropic mesh generator. INRIA-Rocquencourt [Online]. Available at: http://www-rocq.inria.fr/gamma/cdrom/www/bamg/eng.htm.

Hecht, F. and Mohammadi, B. (1997). Mesh adaptation by metric control for multi-scale phenomena and turbulence. In *35th AIAA Aerospace Sciences Meeting and Exhibit*, AIAA-1997-0859, Reno, NV.

Hirsch., C. (1988). *Numerical Computation of Internal and External Flows*. John Wiley & Sons, Chichester.

Huang, W. (2005). Metric tensors for anisotropic mesh generation. *J. Comp. Phys.*, 204(2), 633–665.

Hughes, T.J.R. and Mallet, M. (1986). A new finite element formulations for computational fluid dynamics III. *Comput. Methods Appl. Mech. Eng.*, 58, 329–336.

Hutton, A.G. and Casey, M.V. (2001). Quality and trust in industrial CFD: A European initiative. Paper 2001-0656, Technical report, AIAA.

Jameson, A. (1987). Successes and challenges in computational aerodynamics. AIAA Paper 1987-1184. In *AIAA 8th Computational Fluid Dynamics Conference*, Honolulu, Hawaii.

Jameson, A. (1993). Artificial diffusion, upwind biasing, limiters and their effect on accuracy and multigrid convergence in transonic and hypersonic flows. Paper 93-3359, AIAA.

Jameson, A. and Yoon, S. (1987). Lower-upper implicit schemes with multiple grids for the Euler equations. *AIAA Journal*, 25(7), 929–935.

Jones, W.T., Nielsen, E.J., Park, M.A. (2006). Validation of 3D adjoint based error estimation and mesh adaptation for sonic boom reduction. AIAA Paper 2006-1150.

Knoll, D.A. and Keyes, D.E. (2004). Jacobian-free Newton-Krylov methods: A survey of approaches and applications. *J. Comp. Phys.*, 193(2), 357–397.

Koren, B. (1993). A robust upwind discretisation method for advection diffusion and source terms. *Notes Numer. Fluid Mech.*, 45, 117–138.

Lallemand, M.H., Steve, H., Dervieux, A. (1992). Unstructured multigridding by volume agglomeration: Current status. *Comput. Fluids*, 21, 397–433.

Larrouturou, B. (1989). How to preserve the mass fractions positivity when computing compressible multi-component flows. RR-1080, INRIA.

Laug, P. and Bourochaki, H. (2003). BL2D-V2, mailleur bidimensionnel adaptatif. RR-0275, INRIA.

van Leer, B. (1977). Towards the ultimate conservative difference scheme IV: A new approach to numerical convection. *J. Comp. Phys.*, 23, 276–299.

van Leer, B. (1979). Towards the ultimate conservative difference scheme V: A second order sequel to Godunov's method. *J. Comp. Phys.*, 32, 101–136.

Lesage, A.-C., Allain, O., Dervieux, A. (2007). On level set modelling of bi-fluid capillary flow. *Int. J. Numer. Methods Fluids*, 53(8), 1297–1314.

Li, X., Shephard, M.S., Beal, M.W. (2005). 3D anisotropic mesh adaptation by mesh modification. *Comput. Methods Appl. Mech. Eng.*, 194(48–49), 4915–4950.

Linde, T. and Roe, P. (1998). On multidimensional positively conservative high resolution schemes. Paper 97-2098, AIAA.

Löhner, R. (1989). Adaptive remeshing for transient problems. *Comput. Methods Appl. Mech. Eng.*, 75, 195–214.

Löhner, R. (1990). Three-dimensional fluid-structure interaction using a finite element solver and adaptive remeshing. *Comput. Sys. in Eng.*, 1(2–4), 257–272.

Löhner, R. (2001). *Applied CFD Techniques. An Introduction Based on Finite Element Methods*. John Wiley & Sons, Ltd, New York.

Löhner, R., Morgan, K., Zienkiewicz, O. (1984). The solution of non-linear hyperbolic equation systems by the finite element method. *Int. J. Numer. Meth. Fluids*, 4, 1043–1063.

Loseille, A. (2008). Adaptation de maillage anisotrope 3D multi-échelles et ciblée à une fonctionnelle pour la mécanique des fluides. Application à la prédiction haute-fidélité du bang sonique. PhD Thesis, Université Pierre et Marie Curie, Paris VI.

Loseille, A. and Alauzet, F. (2009). Optimal 3D highly anisotropic mesh adaptation applied to steady flows computation. In *Proc. of the 18th International Meshing Roundtable*, Salt Lake City, UT.

Loseille, A. and Alauzet, F. (2011a). Continuous mesh framework. Part I: Well-posed continuous interpolation error. *SIAM Journal on Numerical Analysis*, 49(1), 38–60.

Loseille, A. and Alauzet, F. (2011b). Continuous mesh framework. Part II: Validations and applications. *SIAM Journal on Numerical Analysis*, 49(1), 61–86.

Loseille, A. and Löhner, R. (2009). On 3D anisotropic local remeshing for surface, volume, and boundary layers. In *Proceedings of the 18th International Meshing Roundtable*, Clark, B.W. (ed.). Springer, Berlin.

Loseille, A., Dervieux, A., Frey, P.J., Alauzet, F. (2007). Achievement of global second-order mesh convergence for discontinuous flows with adapted unstructured meshes. AIAA Paper, 2007-4186.

Loseille, A., Dervieux, A., Alauzet, F. (2010a). A 3D goal-oriented anisotropic mesh adaptation applied to inviscid flows in aeronautics. In *48th AIAA Aerospace Sciences Meeting and Exhibit, AIAA-2010-1067*, Orlando, FL.

Loseille, A., Dervieux, A., Alauzet, F. (2010b). Fully anisotropic goal-oriented mesh adaptation for 3D steady Euler equations. *J. Comput. Phys.*, 229, 2866–2897.

Luo, H., Baum, J., Löhner, R. (1998). A fast, matrix-free implicit method for compressible flows on unstructured grids. *J. Comp. Phys.*, 146, 664–690.

Luo, H., Baum, J.D., Löhner, R. (2001). An accurate, fast, matrix-free implicit method for computing unsteady flows on unstructured grids. *Comput. Fluids*, 30, 137–159.

Magoules, F. (2011). *Computational Fluid Dynamics*. CRC Press, Boca Raton/London/New York/Washington, DC.

Marquant, G., Pateux, S., Labit, C. (2000). Mesh-based scalable image coding with rate-distortion. In *European Signal Processing Conference EUSIPCO 2000*, Tampere.

Martin, R. and Guillard, H. (1996). A second-order defect correction scheme for unsteady problems. *Comput. Fluids*, 25(1), 9–27.

Mavriplis, D.J. (1990). Adaptive mesh generation for viscous flows using Delaunay triangulation. *J. Comp. Phys.*, 90, 271–291.

Mavriplis, D.J. (1997). Unstructured grid techniques. *Annual Review of Fluid Mechanics*, 29, 473–514.

Mer, K. (1998). Variational analysis of a mixed element volume scheme with fourth-order viscosity on general triangulations. *Comput. Methods Appl. Eng.*, 153, 45–62.

Morinishi, K. (2001). Gridless type solver-generalized finite difference method. *Numer. Fluid Mech.*, 78, 43–58.

Morrison, J.H. (2022). 7th AIAA CFD Drag Prediction Workshop [Online]. Available at: https://aiaa-dpw.larc.nasa.gov/.

Murrone, A. and Guillard, H. (2005). A five equation model for compressible two-phase flow computations. *J. Comp. Phys.*, 202(2), 664–698.

Nadler, E.J. (1985). Piecewise linear approximation on triangulations of a planar regions. PhD Thesis, Brown University, Providence, RI.

Naylor, D.J. (1999). Filling space with tetrahedra. *Int. J. Numer. Meth. Eng*, 44, 1383–1395.

Nielsen, E.J. and Anderson, W.K. (2002). Recent improvement in aerodynamic design optimization on unstructured meshes. *AIAA Journal*, 40(6), 1155–1163.

Oden, J.T. (1989). Adaptive methods in computational fluid dynamics. In *Finite Element Analysis in Fluids*, vols 2–10, Chung, G.R. and Abd Karr, T.J. (eds). 7th International Conference on Finite Element in flow problems, Huntsville, Alabama, USA, Taylor and Francis, London.

Olsson E., Kreiss G., Zahedi S. (2007). A conservative level set method for two phase flow II. *J. Comput. Phys.*, 225(1), 785–807.

Osher, S. and Fedkiw, R. (2002). *Level Set Methods and Dynamic Implicit Surfaces*. Springer, Berlin.

Osher, S. and Sethian, J.A. (1988). Fronts propagating with curvature-dependent speed: Algorithms based on Hamilton-Jacobi formulations. *J. Comput. Phys.*, 79(1), 12–49.

Pain, C.C., Humpleby, A.P., de Oliveira, C.R.E., Goddard, A.J.H. (2001). Tetrahedral mesh optimisation and adaptivity for steady-state and transient finite element calculations. *Comput. Methods Appl. Mech. Eng.*, 190, 3771–3796.

Palmerio, B. (1988). A two dimensional FEM adaptative moving node method for steady Euler flow simulation. *Comput. Methods Appl. Mech. Eng.*, 71, 315–340.

Palmerio, B. (1994). An attraction-repulsion mesh adaption model for flow solution on unstructured grids. *Comput. Fluids*, 23(3), 487–506.

Palmerio, B. (1996). Coupling mesh and flow in viscous fluid calculations on unstructured triangular finite element. *Comput. Fluid Dyn.*, 6, 275–290.

Palmerio, B. and Dervieux, A. (1990). On weak and strong coupling between mesh adaptors and flow solvers. In *12th International Conference on Numerical Methods in Fluid Dynamics, University of Oxford, July 9–13, 1990*, Lecture Notes in Physics, vol. 371, Morton, K. (ed.), Springer Verlag.

Palmerio, B. and Dervieux, A. (1996). Multimesh and multiresolution analysis for mesh adaptive interpolation. *Applied Numerical Mathematics*, 22, 477–493.

Peraire, J., Vahdati, M., Morgan, K., Zienkiewicz, O.C. (1987). Adaptive remeshing for compressible flow computations. *J. Comput. Phys.*, 72, 449–466.

Peraire, J., Peiro, J., Morgan, K. (1992). Adaptive remeshing for three-dimensional compressible flow computations. *J. Comput. Phys.*, 103, 269–285.

Perthame, B. and Khobalate, B. (1992). Maximum principle on the entropy and minimal limitations for kinetic scheme. Research report 1628, INRIA.

Perthame, B. and Shu, C.-W. (1996). On positivity preserving finite-volume schemes for Euler equations. *Numerical Mathematics*, 73, 119–130.

Picasso, M. (2003). An anisotropic error indicator based on Zienkiewicz-Zhu error estimator: Application to elliptic and parabolic problems. *SIAM J. Sci. Comp.*, 24(4), 1328–1355.

Piperno, S. (1996). Schéma TVD d'ordre élevé pour la résolution de l'équation de Burgers. Research report 1996-49, CERMICS.

Piperno, S. and Depeyre, S. (1998). Criteria for the design of limiters yielding efficient high resolution TVD schemes. *Comput. Fluids*, 27(2), 183–197.

Popinet, S. and Zaleski, S. (1999). A front-tracking algorithm for accurate representation of surface tension. *Int. J. Numer. Methods Fluids*, 30(6), 775–793.

Quirk, J.J. (1991). An adaptive grid algorithm for computational shock hydrodynamics. PhD Thesis, College of Aeronautics, Cranfield Institute of Technology.

Remacle, J.-F., Li, X., Shephard, M.S., Flaherty, J.E. (2005). Anisotropic adaptive simulation of transient flows using discontinuous Galerkin methods. *Int. J. Numer. Methods Eng*, 62, 899–923.

Rhie, C.M. and Chow, W.L. (1983). Numerical study of the turbulent flow past an airfoil with trailing edge separation. *AIAA Journal*, 21(11), 1525–1532.

Roache, P.J. (1994). Perspectice: A method for uniform reporting of grid refinement studies. *J. Fluid Eng.*, 116(3), 405–413.

Roe, P. (1981). Approximate Riemann solvers, parameter vectors, and difference schemes. *J. Comp. Phys.*, 43, 357–372.

Romberg, W. (1955). Vereinfachte numerische integration. *Kgl. Norske Vid. Selsk. Forsk.*, 28, 7, 30–36.

Rumsey, C.L., Slotnick, J.P., Sclafani, A.J. (2018). Overview and summary of the third AIAA high lift prediction workshop. In *56th AIAA Aerospace and Sciences Meeting*, AIAA-2018-1258, Kissimmee, FL.

Sarkis, M. and Koobus, B. (2000). A scaled and minimum overlap restricted additive Schwarz method with application on aerodynamics. *Comput. Methods Appl. Mech. Eng.*, 184, 391–400.

Saurel, R. and Abgrall, R. (1999). A multiphase Godunov method for compressible multifluid and multiphase flows. *J. Comput. Phys.*, 150(2), 425–467.

Schall, E., Leservoisier, D., Dervieux, A., Koobus, B. (2004). Mesh adaptation as a tool for certified computational aerodynamics. *Int. J. Numer. Methods Fluids*, 45, 179–196.

Selmin, V. and Formaggia, L. (1992). Simulation of hypersonic flows on unstructured grids. *Int. J. Numer. Methods Eng.*, 34, 569–606.

Selmin, V. and Formaggia, L. (1998). Unified construction of finite element and finite volume discretizations for compressible flows. *Int. J. Numer. Methods Eng.*, 39(1), 1–32.

Senechal, M. (1981). Which tetrahedra fill space? *Mathematics Magazine*, 54, 227–243.

Sethian, J.A. (1999). *Level Set Methods and Fast Marching Methods*. Cambridge University Press, Cambridge, MA.

Sharov, D. and Nakahashi, K. (1997). Reordering of hybrid unstructured grids for lower-upper symmetric Gauss-Seidel computations. *AIAA Journal*, 36(1), 484–486.

Sharov, D., Luo, H., Baum, J.D., Löhner, R. (2000). Implementation of unstructured grid GMRES+LU-SGS method on shared-memory, cache-based parallel computers. AIAA Paper, 2000-0927.

Shewchuk, J.R. (2002). What is a good linear element? Interpolation, conditioning and quality measures. In *Proceedings of 11th International Meshing Roundtable*, Ithaca, NY.

Shu, C.W. and Osher, S. (1988). Efficient implementation of essentially non-oscillatory shock-capturing schemes. *J. Comput. Phys.*, 77, 439–471.

Simon, J. (1990). Non-homogenous viscous incompressible fluids: Existence of velocity, density, and pressure. *SIAM J. Math. Anal.*, 21(5), 1093–1117.

Smolianski, A. (2005). Finite-element/level-set/operator-splitting (FELSOS) approach for computing two-fluid unsteady flows with free moving interfaces. *Int. J. Numer. Methods Fluids*, 48, 231–269.

Sommerville, D.Y.M. (1923). Space-filling tetrahedra in Euclidean space. *Proc. Edinburgh Math. Soc.*, 41, 49–57.

Spalart, P.R. and Allmaras, S.R. (1992). A one-equation turbulence model for aerodynamic flows. In *30th AIAA Aerospace Sciences Meeting and Exhibit*, AIAA-92-0439, Reno, NV.

Spalart, P.R. and Rumsey, C.L. (2007). Effective inflow conditions for turbulence models in aerodynamic calculations. *AIAA Journal*, 45(10), 2544–2553.

Spiteri, R.J. and Ruuth, S.J. (2002). A new class of optimal high-order strong-stability-preserving time discretization methods. *SIAM J. Numer. Anal.*, 40(2), 469–491.

Steger, J.L., Dougherty, F.C., Benek, J.A. (1983). A chimera grid scheme. Applied Mechanics, Bioengineering and Fluid Engineering Conference, Houston, TX, USA. In *Advances in Grid Generation*, vol. 5, Ghia, K.N. and Ghia, U. (eds). ASME FED.

Stoufflet, B., Periaux, J., Fezoui, L., Dervieux., A. (1996). 3-D hypersonic Euler numerical simulation around space vehicles using adapted finite elements. In *25th AIAA Aerospace Meeting, Reno (1987)*, AIAA Paper 86-0560.

Sussman, M., Fatemi, E., Smereka, P., Osher, S. (1998). An improved level set method for incompressible two-phase flows. *Comput. Fluids*, 27(5–6), 663–680.

Sykora, S. (1974). Quantum theory and the Bayesian inference problems. *J. Statistical Phys.*, 11(1), 17–27.

Tam, A., Ait-Ali-Yahia, D., Robichaud, M.P., Moore, M., Kozel, V., Habashi, W.G. (2000). Anisotropic mesh adaptation for 3D flows on structured and unstructured grids. *Comput. Methods Appl. Mech. Eng.*, 189, 1205–1230.

Tie, B. and Aubry, D. (1999). Adaptative FE strategy for non-linear and coupled structural computation. In *Numerical Methods in Engineering '96: Proceeding of the Second ECCOMAS Conference on Numerical Methods in Engineering*, Paris, 9–13 September 1996.

Toro, E. (2009). *Riemann Solvers and Numerical Methods for Fluid Dynamics*, 3rd edition. Springer, New York.

Unverdi, S.O. and Tryggvason, G. (1992). A front-tracking method for viscous, incompressible, multi-fluid flows. *J. Comput. Phys.*, 100(1), 25–37.

Vassilevski, Y.V. and Agouzal, A. (2005). Unified asymptotic analysis of interpolations errors for optimal meshes. *Dokl. Math.*, 72(3), 295–298.

Vàzquez, M., Koobus, B., Dervieux, A. (2004). Multilevel optimisation of a supersonic aircraft. *Finite Elem. Anal. Des.*, 40, 2101–2124.

Venditti, D.A. and Darmofal, D.L. (2003). Anisotropic grid adaptation for functional outputs: Application to two-dimensional viscous flows. *J. Comput. Phys.*, 187(1), 22–46.

Venkatakrishnan, V. (1996). A perspective on unstructured grid flow solvers. *AIAA J.*, 34, 533–547.

Venkatakrisnan, V. (1998). Barriers and challenges in CFD. In *ICASE Workshop, ICASE/LaRC Interdisciplinary Series*, vol. 6. Kluwer Academic Publishers, Dordrecht.

Whitaker, D.L., Grossman, B., Löhner, R. (1989). Two-dimensional euler computations on a triangular mesh using an upwind, finite-volume scheme. AIAA Paper, 89-0365.

Wu, H. and Wang, L. (1995). Non-existence of third order accurate semi-discrete MUSCL-type schemes for nonlinear conservation laws and unified construction of high accurate ENO schemes. In *Sixth International Symposium on Computational Fluid Dynamics*, Lake Tahoe, NV.

Zienkiewicz, O.C. and Wu, J. (1994). Automatic directional refinement in adaptive analysis of compressible flows. *Int. J. Numer. Methods Eng*, 37, 2189–2210.

Zienkiewicz, O.C. and Zhu, J.Z. (1992). The superconvergent patch recovery and a posteriori error estimates. Part 1: The recovery technique. *Int. J. Numer. Methods Eng.*, 33(7), 1331–1364.

Index

I, L

M, N

R, S

T, U

Summary of Volume 2

Chapter 5. Goal-Oriented Adaptation for Viscous Steady Flows

Chapter 6. Norm-Oriented Formulations

Other titles from

ISTE

in

Numerical Methods in Engineering

2022

BOUCLIER Robin, HIRSCHLER Thibault
IGA: Non-conforming Coupling and Shape Optimization of Complex Multipatch Structures (Isogeometric Analysis Tools for Optimization Applications in Structural Mechanics Set – Volume 1)

GRANGE Stéphane, SALCIARINI Diana
Deterministic Numerical Modeling of Soil–Structure Interaction

2021

GENTIL Christian, GOUATY Gilles, SOKOLOV Dmitry
Geometric Modeling of Fractal Forms for CAD
(Geometric Modeling and Applications Set – Volume 5)

2020

GEORGE Paul Louis, ALAUZET Frédéric, LOSEILLE Adrien, MARÉCHAL Loïc
Meshing, Geometric Modeling and Numerical Simulation 3: Storage, Visualization and In Memory Strategies
(Geometric Modeling and Applications Set – Volume 4)

SIGRIST Jean-François
Numerical Simulation, An Art of Prediction 2: Examples

2019

DA Daicong
Topology Optimization Design of Heterogeneous Materials and Structures

GEORGE Paul Louis, BOROUCHAKI Houman, ALAUZET Frédéric, LAUG Patrick, LOSEILLE Adrien, MARÉCHAL Loïc
Meshing, Geometric Modeling and Numerical Simulation 2: Metrics, Meshes and Mesh Adaptation
(Geometric Modeling and Applications Set – Volume 2)

MARI Jean-Luc, HÉTROY-WHEELER Franck, SUBSOL Gérard
Geometric and Topological Mesh Feature Extraction for 3D Shape Analysis
(Geometric Modeling and Applications Set – Volume 3)

SIGRIST Jean-François
Numerical Simulation, An Art of Prediction 1: Theory

2017

BOROUCHAKI Houman, GEORGE Paul Louis
Meshing, Geometric Modeling and Numerical Simulation 1: Form Functions, Triangulations and Geometric Modeling
(Geometric Modeling and Applications Set – Volume 1)

2016

KERN Michel
Numerical Methods for Inverse Problems

ZHANG Weihong, WAN Min
Milling Simulation: Metal Milling Mechanics, Dynamics and Clamping Principles

2015

ANDRÉ Damien, CHARLES Jean-Luc, IORDANOFF Ivan
3D Discrete Element Workbench for Highly Dynamic Thermo-mechanical Analysis
(Discrete Element Model and Simulation of Continuous Materials Behavior Set – Volume 3)

JEBAHI Mohamed, ANDRÉ Damien, TERREROS Inigo, IORDANOFF Ivan
Discrete Element Method to Model 3D Continuous Materials
(Discrete Element Model and Simulation of Continuous Materials Behavior Set – Volume 1)

JEBAHI Mohamed, DAU Frédéric, CHARLES Jean-Luc, IORDANOFF Ivan
Discrete-continuum Coupling Method to Simulate Highly Dynamic Multi-scale Problems: Simulation of Laser-induced Damage in Silica Glass
(Discrete Element Model and Simulation of Continuous Materials Behavior Set – Volume 2)

SOUZA DE CURSI Eduardo
Variational Methods for Engineers with Matlab®

2014

BECKERS Benoit, BECKERS Pierre
Reconciliation of Geometry and Perception in Radiation Physics

BERGHEAU Jean-Michel
Thermomechanical Industrial Processes: Modeling and Numerical Simulation

BONNEAU Dominique, FATU Aurelian, SOUCHET Dominique
Hydrodynamic Bearings – Volume 1
Mixed Lubrication in Hydrodynamic Bearings – Volume 2
Thermo-hydrodynamic Lubrication in Hydrodynamic Bearings – Volume 3
Internal Combustion Engine Bearings Lubrication in Hydrodynamic Bearings – Volume 4

DESCAMPS Benoît
Computational Design of Lightweight Structures: Form Finding and Optimization

2013

YASTREBOV Vladislav A.
Numerical Methods in Contact Mechanics

2012

DHATT Gouri, LEFRANÇOIS Emmanuel, TOUZOT Gilbert
Finite Element Method

SAGUET Pierre
Numerical Analysis in Electromagnetics

SAANOUNI Khemais
Damage Mechanics in Metal Forming: Advanced Modeling and Numerical Simulation

2011

CHINESTA Francisco, CESCOTTO Serge, CUETO Elias, LORONG Philippe
Natural Element Method for the Simulation of Structures and Processes

DAVIM Paulo J.
Finite Element Method in Manufacturing Processes

POMMIER Sylvie, GRAVOUIL Anthony, MOËS Nicolas, COMBESCURE Alain
Extended Finite Element Method for Crack Propagation

2010

SOUZA DE CURSI Eduardo, SAMPAIO Rubens
Modeling and Convexity

2008

BERGHEAU Jean-Michel, FORTUNIER Roland
Finite Element Simulation of Heat Transfer

EYMARD Robert
Finite Volumes for Complex Applications V: Problems and Perspectives

FREY Pascal, GEORGE Paul Louis
Mesh Generation: Application to finite elements – 2nd edition

GAY Daniel, GAMBELIN Jacques
Modeling and Dimensioning of Structures

MEUNIER Gérard
The Finite Element Method for Electromagnetic Modeling

2005

BENKHALDOUN Fayssal, OUAZAR Driss, RAGHAY Said
Finite Volumes for Complex Applications IV: Problems and Perspectives

Printed and bound by CPI Group (UK) Ltd, Croydon, CR0 4YY
19/04/2026
14862996-0002